FOR BETTER OR FOR WORSE

ALFRED K. MANN

For Better or for Worse

THE MARRIAGE OF SCIENCE AND GOVERNMENT IN THE UNITED STATES

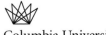

Columbia University Press *New York*

Columbia University Press
Publishers Since 1893
New York   Chichester, West Sussex

Copyright © 2000 Columbia University Press
All rights reserved

Library of Congress Cataloging-in-Publication Data
Mann, Alfred K.
For better or for worse : the marriage of science and
government in the United States / Afred K. Mann
p. cm.
Includes bibliographical references and index.
ISBN 0–231–11706–X (cloth: alk. paper)
1. Science and state—United States—History—20th century.
I. Title
Q127.U5 M36 2000
509.73—dc21   00–060120

Printed in the United States of America
Designed by Audrey Smith

c 10 9 8 7 6 5 4 3 2 1

To my wife, Jayne, who did not live to see this book published, and to our children, Stephen, Cecile, David, and Brian, who helped me to see it through to the end.

Contents

List of Illustrations

List of Tables

Early in the twentieth century, when funds from wealthy individuals and private foundations ceased to meet the needs of modern science in the United States, the federal government began to invest in a national scientific infrastructure. This was done tentatively at first and then in World War II on the largest scale imaginable. The investment was so successful that it virtually demanded to be continued when peace came. So began the development of an American science establishment, today an amalgam of scientists, engineers, universities, industrial laboratories, and federal science agencies. The establishment is a remarkable achievement in its own right, distinct from the science and technology it has helped to produce but an integral part of them. It has been held together for a half century by a federal government determined to foster the benefits of science and technology for its citizens. The government has achieved this using public money to underwrite the cost of the science establishment despite the intrinsic fluidity and ungovernable nature of both the science and the establishment.

By chance, my career coincided with the emergence and growth of the science establishment. In the words of Dean Acheson, I was "present at the creation." My early impression of the establishment as a loose patchwork of federal agencies and private institutions underwent a significant change as time passed. I once thought them to be independent fiefdoms connected only by a common interest in the federal budget. That perception was replaced by an awareness of their joint dedication to encouraging and supporting science and technology for the benefit of the nation.

The science establishment is not usually acknowledged as a separate entity in what is written about science and technology. I hope to compensate for that omission in this book, which is an overview of the science establishment and its relationship with the federal government. I have traced the development of the four nonmilitary federal science agencies that have been and still are the principal supporters of basic scientific research and technology in U.S. universities, where most of the fundamental research in the nation takes place. I believe that the essential features of the science establishment as a whole appear clearly in this description of the evolution of the individual federal science agencies.

This book is not a scholarly history with any claim to completeness. Instead, it attempts to tell the story of the complex relationship between science and government in the United States as one might tell the story of a marriage between two people. This analogy is not, I think, too finely drawn and helps to make the changing fortunes within the union easier to follow.

It is a pleasure to acknowledge the kindness of individuals who suggested reference material and in many instances furnished it to me. These were the historians: George Mazuzan at the National Science Foundation, Victoria Harden and Sam Josaloff at the National Institutes of Health, Marie Hallion at the Department of Energy, and Stephen J. Garber at the National Aeronautics and Space Administration. In addition, D. Allan Bromley, Richard Mandel, and Sanford P. Markey encouraged me and sent valuable material.

The science editor at Columbia University Press, Holly Hodder, her assistant, Jonathan Slutsky, and the copyeditor, Sarah St. Onge, eliminated occasional awkwardness in my presentation and corrected egregious errors with tact and forbearance. With a sharp eye for mistakes and ever-present good humor, Jean O'Boyle has typed the many drafts it has taken to achieve a finished manuscript.

FOR BETTER OR FOR WORSE

CHAPTER 1

Introduction

When Franklin Delano Roosevelt and a Democratic congress won control of the federal government of the United States in 1932, a controversial new view of how federal influence and money might be used to change American society was also inaugurated. The active role of the federal government in pursuit of that view led to the creation of many federal agencies in the quarter century between 1932 and 1957 and brought about a very different nation than had existed previously. With the approval of the American public, that view is still firmly in place sixty-five years later, but controversy continues over how widely it should be extended and precisely how it should be implemented.

One far-reaching change that came about during World War II was an increase in government involvement in scientific research. This area was largely left to the private sector of the United States before the war, apart from certain special developments that were the result of patient, determined pressure from forward-looking private citizens. For example, soon after the first powered airplane flight at Kitty Hawk, the enormous potential of aviation required an agency to coordinate research and development and advise the government on progress in aviation. After years of prodding—remember that the Wright brothers successfully flew their airplane in 1903 and made trial flights for the War Department in 1908—the National Advisory Committee for Aeronautics was created in 1915, one year after the beginning of the First World War. It is worth noting that NACA, as it came to be called, received an annual budget of $5,000 for its first five years. By the early 1920s those funds were supplemented by private money from the Daniel Guggenheim Fund for the Promotion of Aeronautics, which supported programs in the new science at several universities and helped to stimulate wide interest in it. By 1929 fourteen hundred aeroengineering students were enrolled in more than a dozen universities in the United States; and in 1930 the Guggenheim Fund helped bring the brilliant Theodore von Karman from Germany to the California Institute of Technology (Caltech), thereby enriching aerodynamic theory and planting the seed from which the famous Jet Propulsion Laboratory of Caltech grew. The influence of NACA will reappear in these pages, evident not only in the development of aeronautics but also as an early training ground for U.S. science policy makers. Apart from NACA, however, and the National Bureau of Standards (NBS), whose name defined its function, virtually all other science laboratories in the nation were in the private sector, mostly in universities.

With the onset of WWII in Europe in 1939, the government and a number of eminent advisers on science recognized that coordination of research and development in areas other than flight and physical and chemical standards—particularly in weapons development and medicine—was a matter of extreme urgency and required government action. How this coordination was carried out and—because it was so successful—its outgrowth after the war are matters of more than historical interest. The idea that emerged from the wartime experience and the decade thereafter was that federal funding for peacetime research in government agencies and institutions outside the government, especially universities, would be vitally important to the well-being of the nation. This idea is no longer subject to serious dissent, but questions of how funding should be provided, how it should be controlled and to what extent, what fraction of the available resources should be devoted to research, and how to set the emphasis between science and technology are still far from being completely resolved after fifty years of government sponsorship of research.

The first step toward the present system occurred near the end of World War II, when on November 17, 1944, President Roosevelt wrote to Vannevar

Bush, then director of the Office of Scientific Research and Development, asking for recommendations of ways in which the federal government might help to promote science and science education after the war. The Office of Scientific Research and Development (OSRD) was a World War II agency that had coordinated development of radar and the proximity fuze, among other important weaponry innovations, and of startling advances in military medicine, all at the same time and in parallel with the development of the atomic bomb by the Manhattan Project. These national efforts are acknowledged to have been eminently successful in coordinating scientific research and applying existing knowledge to the solution of the technical problems that arose in the war.

President Roosevelt's letter was a model of its kind, brief but specific in its request, optimistic and uplifting, as one of its paragraphs conveys: "New frontiers of the mind are before us, and if they are pioneered with the same vision, boldness, and drive with which we have waged this war we can create a fuller and more fruitful employment and a fuller and more fruitful life." By the time Vannevar Bush was able to reply in July 1945, however, President Roosevelt was dead, and it was to the new president, Harry S. Truman, that Bush transmitted his report, titled *Science: The Endless Frontier*. The report laid out a detailed plan for the federal support of science and science education in peacetime, much of which was implemented in the Truman and Eisenhower administrations. It was effectively responsible for the creation of two government science agencies: directly for the National Science Foundation and indirectly for the growth of the National Institutes of Health. Of equal importance, it emphasized the code of behavior to be followed by those agencies and other federal science agencies in carrying out the goals outlined in President Roosevelt's letter.

Despite its remarkable influence on successive administrations and the way it has contributed to the progress of American science during the fifty years since it was written, many scientists and certainly the general public are largely unaware of the content of *Science: The Endless Frontier*. Yet there is still much to be learned from the report, which incidentally is an easy read without legal or scientific jargon. In the United States today, where the yearly federal investment in basic science and technology—in what we might call the science establishment—amounts to about thirty-five billion dollars, it is valuable to recognize the roots of that establishment (the OSRD, the Manhattan Project, and the Bush report) and to follow the evolution of the system by which the federal government now funds and encourages science. This enables us to ask to what extent the system has been successful and whether it now needs to be changed in part or in whole. There is special urgency to answering this question today, in the wake of the cold war, when the issue of federal government support of research confronts the need for greater protection of the environment and increased social services, as well as the demand to balance the federal budget.

The heart of the present U.S. science establishment consists of five major federal science agencies: the National Institutes of Health (NIH) in the Department of Health and Human Services (DHHS), the Department of Defense (DOD), the National Aeronautics and Space Administration (NASA), the Department of Energy (DOE), and the National Science Foundation (NSF). These agencies are responsible for 85 percent of the federal government's basic science and technology budget; the remaining 15 percent is spent by about fifteen other federal agencies for which science and technology are peripheral to their main function. It is important to recognize that all the federal funding of science in colleges and universities flows through these agencies. This book will examine the four nonmilitary (non-DOD) agencies that affect science in the colleges and universities and describe how they got to be what they are today. The level of support made available by the DOD for basic science and technology in U.S. colleges and universities is unclear, reflecting Congress's ambivalent attitude toward the DOD's involvement in educational institutions, even in circumstances that might be beneficial to both parties. As a consequence, the DOD is not recognized as a major funder of basic scientific research in colleges and universities.

In attempting to justify the federal government's continuing investment in science and technology, scientists and engineers often point to particular achievements, a tactic that at best is only partly convincing and usually succeeds in pitting members of one scientific discipline against members of another. In fact the best way to argue for extensively funding science and technology is simply to say: Look around you, and don't get hung up on the occasional mistakes, conflicting claims and false advertising, or abuse of scientific knowledge. Instead, pay attention to the interplay between modern science and technology that has given rise to deeper insights into the world around us, while at the same time improving the physical comfort and quality of our lives. Astronomy, chemistry, and physics—the so-called hard sciences—have advanced in ways that could not be foreseen at the turn of the twentieth century and continue to move forward with unabated momentum. The biological sciences have progressed from the narrow paths of descriptive disciplines preoccupied with categorization and nomenclature to the wider boulevard of molecular and cell biologies. Computer science and computers have invaded most households and businesses as well as science laboratories. In the industrially developed nations at least, human life expectancy has been substantially increased—for example, from the fifty-year average life expectancy in the United States in the early 1950s to the remarkable life expectancy of seventy-six years in 1993. At the same time, much human pain and suffering have been alleviated by advances in medicine, dentistry, and personal and public hygiene, all products of the better understanding brought about by science and technology.

Taken together, these achievements sustain each other across disciplinary boundaries through new ideas and new experimental techniques. But the sustenance common to them all has been financial support by the federal government during the past half century, which in turn stems from public approval of that use of tax money. The American science establishment is a vast enterprise, however, and its public approval depends on the degree to which the significance and vitality of the enterprise are made clear not only by media reports of new accomplishments but also by periodic overviews of the enterprise as a whole. A parallel purpose of this book is to provide one such overview.

The next chapter describes the OSRD and the Manhattan Project and their accomplishments in WWII. This is primarily to emphasize the spirit and style they established. A short account of Vannevar Bush, the unique individual who was the first presidential science adviser and the architect of the present science establishment, is included for the same reason. Chapters 3 through 7 recount the highs and lows of the fortunes of the science establishment, decade by decade, from the tumultuous beginning of each science funding agency through the changes forced on them, individually and jointly, by national and world events, up to the present. The last chapter is a summary and an attempt to look into the future of science not only as an individual pursuit but also as an organized, societal purpose in the hands of successive new generations.

CHAPTER 2

Love at First Sight: 1939–1945

The Office of Scientific Research and Development was organized to coordinate weapons development and medical care for the U.S. military as World War II approached.

The Office of Scientific Research and Development (OSRD) was the product of the leaders of the U.S. scientific community, among them Vannevar Bush, former vice president of engineering at the Massachusetts Institute of Technology, head of the Carnegie Institution of Washington, and director of the National Defense Research Committee (NDRC). Bush's principal colleagues were James B. Conant, a distinguished chemist and president of Harvard University; Frank B. Jewett, director of Bell Telephone Laboratories and president of the U.S. National Academy of Sciences; and Karl T. Compton, president of MIT. A. N. Richards, vice president for medical affairs of the University of Pennsylvania, was chairman of the OSRD's Committee on Medical Research. These individuals had earned the trust and respect of American scientists, but, most important, they had the support of the White House, from which the official and financial strength of the OSRD would eventually come. Together, they created a centralized, civilian research organization that would develop weapons and improved medical care for the war they believed that the United States would probably enter.

The NDRC represented the first attempt to put U.S. science and technology on a wartime footing. At the urging of Vannevar Bush in June 1940, President Roosevelt created the NDRC—his personal advisory committee for wartime science and technology—composed of civilian scientists who maintained a broader perspective than military advisers did. Although the NDRC could draw from its own funds to perform its assignments, and although Bush had direct access to the president, the NDRC was not comprehensive enough to translate its recommendations into action. In the summer of 1941, urged once again by Bush, Roosevelt created the OSRD and assigned Bush to coordinate all scientific and technological actions connected with the national preparation for defense in a world at war.

The OSRD was organized loosely on purpose: this encouraged initiative and innovation on the part of working scientists and engineers. Perhaps most significant was that the OSRD opted not to construct or operate laboratories of its own. Instead, it turned largely to universities, especially those that had previously demonstrated the capability to carry out broad-ranging assignments. The OSRD broke new ground by contracting universities on a cost basis that included expenses for universities' overhead. These contracts were drawn without the restriction to safeguard against profiteering ordinarily included in most contracts during WWII. The contracts were so successful—no mismanagement was found in later surveys—that they became the basis for federal funding of university scientific research after the war.

The flexible organization developed by the OSRD was able to merge young military personnel—often with battle experience—with civilian scientists and soon earned the respect of the senior officers responsible for weapons and weapon development in the armed services. Without the complete cooperation of those officers, advances in weapons and medicine emerging from the OSRD might have been delayed or even prevented from reaching the battle areas. At the same time, Bush and his colleagues knew the limits of civilian interference with the military and scrupulously stayed back from those limits. The smooth, productive way in which the largely civilian OSRD joined with and was accepted by the military in WWII was unprecedented. It has been a model for cooperation between civilian scientists and the military ever since.

Soon after the end of World War II, a series of books available to the general public under the title *Science in World War II* described the organization and accomplishments of the OSRD, which had been secret during the war. This series, consisting of seven parts contained in more than twenty volumes, explains in detail the areas in which the OSRD functioned and the breadth of its contributions to the science and technology of modern war. The titles of the seven parts of *Science in World War II* make clear that the work of the OSRD covered the entire allied war effort in science, technology, and medicine, excepting only the atomic bomb project: "New Weapons of Air Warfare," "Combat Scientists," "Advances in Military Medicine," "Rockets, Guns, and Targets," "Chemistry," "Applied Physics: Electronics; Optics; Metallurgy," and "Organizing Scientific Research for War." All were written by the people who did the work. There is also an official history of the OSRD—*Scientists Against Time*, by James Phinney Baxter III, with an introduction by Vannevar Bush—which was published in 1946.

The OSRD acted as stimulus, research agency, and mediator between civilian scientists and the military to bring about many important developments in weaponry.

Examples of the scope and variety of OSRD achievements are as compelling as they are significant. One of them, radar (radio detection and ranging) became a marvelously versatile instrument of warfare and transportation safety, revolutionizing both those areas. A radar unit broadcasts a radio beam focused by a parabolic reflector, like the light from an automobile headlight. A receiver in the unit picks up a portion of the broadcasted beam reflected back by the object or target it strikes. By measuring the total time required for the radio beam to travel to the target and return to the unit, the distance or range to the target is determined. The direction in which the beam is aimed locates the position of the target in space. The picture of an area scanned by radar, as described by one of the pioneers of radar development, Sir Robert Watson-Watt, "is a map-like outline in which seas, lakes and waterways remain black. . . . Coastlines with their cliffs, bays and inlets show up clearly as outline map features because they scatter radiation back to its source. . . . The inland landscape is a nondescript intermediate tone; and . . . 'the works of man'—camps, hangars and above all towns and cities—stand out brightly."[1] For locating the range and direction of distant aircraft, the efficiency of radar was unparalleled.

As early as 1925, two scientists at the Carnegie Institution of Washington, Gregory Breit and Merle A. Tuve, used pulses of radio waves to measure the height of the ionosphere, the layer of ionized air at high altitude that reflects radio waves back to Earth. This method of sending out a train of radio pulses and timing the echo became the standard for study of the ionosphere the world over. The idea of using the technique to detect aircraft and ships soon was taken up by scientists in the United States, England, France, and Germany. By the end of 1935, stimulated by fear of German air attack, the British constructed a chain of five stations for the radio location of aircraft on the east coast of England. The United States was not far behind and installed radio-locating units on its battleships *New York* and *Texas* in 1938 and on the carrier *Yorktown* in 1940. A radar unit developed by the U.S. Army detected Japanese aircraft approaching Pearl Harbor, but the observation was disregarded and the device turned off because it was thought to be malfunctioning.

The major advance in radar technology, however, came with the discovery in 1940 of a revolutionary radio wave generator, the resonant cavity magnetron, capable of producing extraordinarily large microwave power for high-resolution observation of very distant targets. The resonant cavity magnetron was the product of British physicists led by M. L. Oliphant, a professor at the University of Birmingham. The prototype, small enough to be carried in a briefcase, was brought to the United States by a British scientific mission, so the story goes, with the magnetron wrapped in newspaper and hand-delivered by Oliphant.

The advent of the magnetron made possible the design and construction of microwave radar units for ships, planes, and tanks—for all the armed services—units that provided enormous range and high resolution. The NDRC, already deeply immersed in radar research and development at the Radiation Lab of MIT and soon to come under the aegis of the OSRD, undertook and completed this task. This new superior radar used by the armed forces of Britain and the United States tipped the balance of the war on land, on sea, and in the air in their favor. By the end of the war, three billion dollars' worth of radar units designed and engineered in the MIT Radar Laboratory (Rad Lab) had been produced in industry and delivered to the armed services, as well as twenty-five million dollars' worth of radar gear supplied directly by the Rad Lab. This involved a hundred and fifty separate and distinct radar systems varying in size from lightweight compact sets for fighter planes and PT boats to the Microwave Early Warning system, housed in five trucks and manned by a company of soldiers. Apart from their use by armed services everywhere, the descendants of these systems are used today in every major airport and airliner in the world, and all but the smallest seagoing vessels rely on them for their safety.

A second example of an OSRD achievement, less widely known by the public then and now, was the radio proximity fuze. This development was a solution to the problem of how to damage or destroy a target that was not directly struck by a projectile—say, an antiaircraft shell—but where instead there was a near miss, that is, within the explosive pattern of the projectile fragments of the shell. The radio proximity fuze was a device closely related in concept to the idea at the heart of radar. Each device broadcast radio pulses, some of which would be reflected from a target to a receiver also in the device, and the received information would provoke action by the device itself. It would be displayed as an active, up-to-date map in a radar set or would trigger the explosion of the antiaircraft shell in which it was mounted, if the strength of signal reflected from the target indicated that the shell was within near-miss distance.

While the idea for the radio proximity fuze was simple and straightforward, implementation of the idea was not. The fuze had to be small enough to fit into a projectile, if possible a five-inch shell, leaving most of the room for explosive, and its components so ruggedly built that they would withstand the accelerations and shocks of being fired from a cannon. The requirements for miniaturization and resistance to mechanical shock had prevented earlier successful manufacture of a proximity fuze, although many patents had been issued in different countries. The Germans, for example, had at various times worked on more than thirty designs and were working on a dozen as late as 1944.

Problems with the design and manufacture of proximity fuzes had been recognized by the U.S. Navy by 1940 (and before that by the British Royal Navy). In that year, the NDRC got involved. At first, both the Americans and the British concentrated on proximity fuzes of a type that might be incorporated in bombs or rockets, because the space limitations in those projectiles were not as restrictive as in artillery shells. But a little later, the NDRC, by then a section within the OSRD, moved toward a radio proximity fuze for antiaircraft shells. Intensive research indicated that rugged, miniature electronic tubes and other components of a radio set capable of sending and receiving its own signal when reflected from a target aircraft were not as difficult to design and build as had been thought originally. Under pressure from the navy, development work at several laboratories—initially the Carnegie Institution of Washington and the National Bureau of Standards and later at the newly formed Applied Physics Laboratory of the Johns Hopkins University—went forward rapidly to test whether the components of the fuze would survive the firing of a shell and whether the necessary safety devices in the fuze to ensure against early misfirings were satisfactory. Finally, shells equipped with radio proximity fuzes had to demonstrate a satisfactory level of efficiency in shooting down aircraft.

Within two years, the fuze was extensively tested in the field in a variety of shells and demonstrated to be highly efficient against aircraft and safe for the personnel who used it. Soon after Pearl Harbor, the fuze was put into full-scale production. At the insistence of the U.S. Navy—a remarkable vote of confidence—the OSRD undertook arrangements for production and quality control. At the height of production there were three hundred companies and two thousand different plants at work on the radio proximity fuze, and nearly two million fuzes were manufactured each month.

Prior to WWII it was estimated that the antiaircraft fire from even the most efficient battery required more than two thousand rounds to bring down one of several attacking planes. The inaccuracy was more the result of shells' failure to explode at the correct range than of poor aim. The range at which a shell would burst was preset by hand in time-fuzed shells just before they were fired. A time-fuzed shell might explode anywhere along a thousand-foot length of its path as a result of the uncertainties inherent in the time-fuzed shells themselves, even if the range to the target had been estimated correctly (which was very hard to do for a rapidly moving airplane). The radio proximity fuze eliminated that defensive weakness just when the needs of the United States and Britain were most desperate. German dive bombers had demonstrated unequivocally in attacks on Belgium, Holland, and France that airpower was to be a dominant factor in WWII. The German campaign in Norway taught a lesson to the British about the ability of the airplane to neutralize seapower, as did the Japanese campaign in the Far East. The security of army ground forces as well as the mobility of fleets were threatened as never before.

The first large-scale use of the fuze was by the U.S. Navy in the Pacific. Improved gun directors—devices driven by a radar unit that automatically aimed and found the range for gun batteries—had somewhat increased antiaircraft battery efficiency but much remained to be done by the radio proximity fuze. Bush, a conservative engineer, estimated that the proximity fuze increased the effectiveness of naval five-inch gun batteries by a factor of seven. Put another way, this was the same as having seven times as many five-inch batteries on a ship.

Another defensive application of the fuze was in combating the first so-called flying bomb, the V-1. This is an extraordinary saga in its own right. In late 1943 secret intelligence indicated that the Germans were setting up to launch robot bombs, the V-1s, against London and southern England, where the Allied invasion force would be gathered in preparation for the invasion of France in mid-1944. It was absolutely imperative to find an effective countermeasure to that threat, and months before the first of the V-1 bombs were launched, intelligence services and the OSRD moved to do so.

The V-1 traveled at three hundred fifty miles per hour, about as fast as fighter planes of the time. The available defensive measures against it included fighter planes that would meet the robots at as great a distance from Britain as possible and, mostly as a last resort, antiaircraft fire at the coasts. When the bombs were first launched against London in June 1944, interceptor aircraft carried the burden of defense. A number of V-1s were shot down, but the toll of life and property destroyed by the bombs that penetrated the interceptor shield was large, and the implication for Allied support of the recently mounted Allied invasion of France was ominous. In the few months between the arrival of the first shipment of proximity fuzes in England and the most intense of the V-1 attacks, however, the British and Americans set up a number of defensive units on the coast of the English Channel that were ready by the second week of July. Each unit consisted of an advanced technology radar installation that fed a high-precision computerized electromechanical system that directed a battery of guns equipped with proximity-fuzed shells.

The V-1 attacks lasted eighty days. During the last four weeks of that period, destruction of V-1s by proximity-fuzed antiaircraft fire steadily grew. In the first week, 24 percent of the targets engaged were destroyed by ground fire using proximity-fuzed shells; in the second week, 46 percent; in the third, 67 percent; and in the last week, 79 percent. The V-1 ceased to be a serious threat.

The system of radar-computerized gun directors and radio proximity fuzes dealt a deadly blow to German forces in still another way. Authorization to use proximity fuzes over land was delayed by the U.S. Joint Chiefs of Staff's concern that an unexploded dud might be retrieved and copied by the enemy. The Joint Chiefs were ultimately persuaded that the time between capture of a dud and the availability of a manufactured supply of fuzes would be long, possibly two years, given the situation in Germany at end of October 1944. After the defeat of the V-1s, the proximity fuze was thus released for use over land, again mostly in conjunction with radar and automated gun directors. The fuze went into use against ground troops and armor at the time of the Battle of the Bulge, the last major German offensive of the war, which had been planned to take place in a period of bad weather when Allied planes could not perform effectively. Extensive studies of how to trigger radio proximity fuzes to detonate howitzer shells at varying heights above land targets had shown them to be more than five times as effective as contact fuzes in the same circumstances. The result was that German divisions, massed in the open and feeling secure against both Allied aircraft and the usual artillery fire, were decimated by proximity-fuzed artillery shells employed efficiently both day and night. The effect was summed up in a letter from General George Patton to General Levin Campbell, Chief of Ordnance, on December 29: "The new shell with the funny fuse is devastating. The other night we caught a German battalion, which was trying to get across the Sauer River, with a battalion concentration and killed by actual count 702. I think that when all armies get this skill we will have to devise some new method of warfare. I am glad that you all thought of it first."[2]

Today, artillery shells and bombs and rockets have evolved far beyond the technology of WWII, but the proximity fuze was a breakthrough in the same sense that radar was a breakthrough. Both revolutionized warfare during WWII and for the next half century. Begun somewhat earlier under the British and the NDRC and perfected and manufactured under the aegis of the OSRD, the work on radar was carried out at the Massachusetts Institute of Technology (MIT) Radiation Laboratory at about the same time that the proximity fuze was developed at the Johns Hopkins University Applied Physics Laboratory. These and other university laboratories, newly staffed with university scientists and engineers called to the war effort by the manpower commission of the time, reflected the coordination of civilian and military talent and resources imposed by the OSRD.

FIGURE 2.1. *Top*: National Defense Research Committee. *Front row, left to right*: F. B. Jewett, president of the National Academy of Sciences; Rear Admiral J. A. Furer, U.S. Navy, coordinator of research and development, Navy Department; J. B. Conant, president of Harvard University; R. C. Tolman, dean of the Graduate School, California Institute of Technology. *Rear row, left to right*: K. T. Compton, president of Massachusetts Institute of Technology; Roger Adams, head of the chemistry department, University of Illinois; C. P. Coe, U.S. commissioner of patents; Irvin Stewart, executive secretary of the Office of Scientific Research and Development.

Source: James Phinney Baxter III, *Scientists Against Time* (Boston: Little, Brown, 1946), p. 15.

Bottom: The Committee on Medical Research, Office of Scientific Research and Development. *Left to right:* Dr. R. E. Dyer, Public Health Service; Rear Admiral Harold W. Smith, U.S. Navy; Dr. A. Baird Hastings; Dr. Chester S. Keefer, medical administrative officer; Dr. A. N. Richards, chairman; Dr. Lewis H. Weed, vice-chairman; Brigadier General James S. Simmons, U.S. Army; Dr. A. R. Dochez; Dr. Irvin Stewart, executive secretary.

Source: James Phinney Baxter III, *Scientists Against Time* (Boston: Little, Brown, 1946), p. 124.

Another achievement of scientists in the OSRD, *the result of medical research on blood, illustrates the variety of the agency's research and the wisdom of its organization.*

Casualties involving hemorrhage, burns, and shock were known before WWII to require immediate treatment to restore blood volume to normal levels. Treatment of victims of auto accidents and other violence in peacetime had made this clear. Before WWII, however, restoring blood to a traumatized person meant having a donor present so that blood could be pumped directly from the donor to the recipient under sterile conditions. This was not possible on the battlefield, and the armed services encouraged the medical and chemical communities to find ways in which blood and blood derivatives might be stored for long periods and made available on demand. Today, the use of stored blood and blood plasma and its components in hospital operating and emergency rooms is commonplace, but in 1940 neither whole blood, nor plasma, nor its derivatives were available. Prompted by the navy, the Department of Physical Chemistry at Harvard University set out to find a remedy and soon was working in conjunction with other medical institutions and the American Red Cross, all coordinated by the Committee on Medical Research (CMR), a section of the OSRD.

It was not by chance that the physical chemists at Harvard were selected for this purpose. They had been studying the physical and chemical properties of proteins, including blood proteins, for more than twenty years and were well known for their accomplishments. It was quickly recognized that in certain circumstances components of blood other than red blood cells were especially valuable. For example, albumin, constituting 60 percent of blood plasma proteins and of small molecular size, was expected to draw fluid back into the blood vessels from the tissues and increase blood volume in a traumatized circulatory system. In extensive clinical testing done at CMR institutions during the winter of 1941–1942, this hypothesis was proved to be correct. Albumin was completely stable and nontoxic at all temperatures, and measurements made of plasma volume in patients before and after injection demonstrated its effectiveness. Human albumin, in concentrations of twenty-five grams of dissolved albumin in one-fifth of a pint of water—more than the amount of albumin in five times that volume of plasma—was formally approved for distribution to the military for transfusions in February 1942.

The attack on Pearl Harbor on December 7, 1941, came just before that formal recommendation. Only a small quantity of processed albumin remained at the time; most of it had been used in the successful clinical tests. But the navy was determined to do its best for the burned and wounded at Pearl Harbor. A telephone call from A. N. Richards, chairman of the CMR, to Edwin Cohn at Harvard set in motion the "as-soon-as-possible" shipment of the last of the albumin supply to the senior medical officer at Pearl Harbor. The results confirmed all the earlier clinical tests: burn victims and those suffering from loss of blood for myriad other reasons, many in deep shock, were rejuvenated and given at the least a temporary lease on life during which additional treatment could be provided.

There is much more to this story of how blood and blood derivatives were made available for immediate, on-the-spot transfusions. The American Red Cross supervised nationwide blood collection. Researchers discovered how to isolate and store other blood components, including whole blood itself, and how to manufacture—while scrupulously preserving sterility—enormous quantities of blood and blood plasma. Thirteen million pints of blood were collected, processed, and distributed to American armed forces everywhere by the end of WWII. And ever since, blood transfusions by the methods first developed during the war have been an invaluable weapon in the physician's arsenal.

Descriptions of these specific accomplishments of the Committee on Medical Research of the OSRD illustrate the magnitude of the challenges it accepted and successfully met. While the problem of blood was being solved, the CMR was also seeing to the production and distribution of sulfa drugs and penicillin on a worldwide scale. At the beginning of the war, there was not enough penicillin in the United States—or anywhere else—to treat a single patient. Imagine the labor and organization required to produce supplies adequate for the entire armed services. These medical products, antibiotics, blood derivatives, and whole blood furnished the life-sustaining support that made the extraordinary advances in WWII surgery possible. Two statistics sum it up: in the U.S. Army, including overseas forces in WWI, the death rate from all diseases was 14.1 per 1,000; in WWII, it was reduced to 0.6 per 1,000. Furthermore, despite the devastating antipersonnel weapons of WWII, the fatality rate among the wounded was lower than in any war in history.

The other major organization for science and technology in WWII was the Manhattan Project, the single purpose of which was to produce an atomic bomb.

"The Manhattan Project" was the popular name given later in the war to an organization very different from the OSRD, although its origins were similar. It seems likely that the term "Manhattan" was originally suggested because the earliest U.S. work on fission was at Columbia University, in Manhattan, New York. Later, when the army became involved, its first project office was also in Manhattan. In both cases, the OSRD and the Manhattan Project, scientists and educators brought to the government's attention scientific problems relating to the situation in Europe and urged the need for U.S. action to prepare to meet them. Indeed, the OSRD and the Manhattan Project were initially joined at the hip: the prospects leading to the Manhattan Project had been surveyed and assessed in studies by committees of scientists and engineers assembled by the OSRD. The OSRD, however, undertook and oversaw a multitude of diverse technical and medical problems, while the Manhattan Project was aimed at one vital goal: the production of an atomic bomb.

Before WWII, the heaviest stable element in the periodic table of the elements was uranium, which is found in ores deposited in several places on the planet. Experiments done in 1938 in Europe, mainly in Germany, indicated, however, that the uranium nucleus was much closer to the brink of instability than anyone had thought. These experiments, which took place a few years after the Nazis came to power but were unrelated to that political change, showed that the uranium nucleus could be induced to break into two lighter nuclei, in the process releasing a very large amount of energy. The reaction, which was given the name "fission" (breaking into two pieces), appeared to have been initiated when the uranium nucleus captured and absorbed a free neutron, the uncharged constituent of atomic nuclei. The early data came as a complete surprise but were soon confirmed and correctly interpreted during a period of great excitement among nuclear scientists in Europe, Britain, and the United States. The possibility of using the discovery to make a bomb of unparalleled destructive power was obvious and became a matter of grave concern within the scientific community. It led to the famous letter from Albert Einstein to President Roosevelt alerting him to the prospect some months after WWII had begun in Europe. Einstein's scientific reputation was unparalleled. A warning from him would not be ignored. He mentioned the work of Enrico Fermi and Leo Szilard in the United States and the more recent work of Frédéric Joliot in France as evidence that a nuclear chain reaction in uranium would be achieved in the immediate future. This new phenomenon, he said, could lead to extremely powerful bombs. A single bomb of the new type exploded in a port might very well destroy the whole port together with some of the surrounding territory. He suggested that an individual be chosen to maintain permanent contact between the administration and the group of physicists working on chain reactions in the United States. Finally, he noted that Germany had stopped the sale of uranium from the Czechoslovakian mines that it had taken over. He speculated that such early action might be explained by the fact that the son of the German under-secretary of state, von Weiszächer, was attached to the Kaiser-Wilhelm Institute in Berlin, where some of the American work on uranium was being repeated.

It was no easy task to convey to the president and other high-ranking government officials the nature of the threat posed by an atomic bomb, partly because it involved unfamiliar scientific ideas and partly because the scientists themselves did not know how real the threat was. The observation of fission in small samples of uranium in the laboratory was clear, and the release of a large amount of energy in each fission reaction was also clear. But whether it was at all possible to create a fission bomb and precisely how to do it were completely unanswered questions. One thing was certain, however: a bomb with the strength equivalent to tens of thousands of tons of TNT in the hands of Nazi Germany alone would be a catastrophe of the first magnitude for the United States and especially for Britain.

Einstein's letter was hand-delivered to President Roosevelt on October 11, 1939, in a meeting organized to brief the president and suggest action he might take. As a result, the president created an advisory committee—guilelessly named the Uranium Committee—to report to him; this represented the first action by the government toward construction of an atomic bomb. The advisory committee, chaired by Lyman J. Briggs, the director of the National Bureau of Standards, met soon after, on October 21, and Briggs reported to the president on November 1 that if the latent energy of fission could be developed, "it might supply power for submarines, and possibly provide . . . a source of bombs with destructiveness much greater than anything now known."[3] The committee report urged the government to support a thorough experimental study of those questions and to purchase several tons of pure graphite and fifty tons of uranium oxide immediately, just in case the preliminary experiments proved promising. It also urged the government to support and coordinate the work in different universities already engaged in fission-related studies and recommended that the Uranium Committee be enlarged to include individuals with a wider spectrum of skills and talents outside the government.

The next few months saw little progress in Washington, although the army and navy did transfer six thousand dollars to the Uranium Committee to purchase materials for experiments on the neutron absorption properties of graphite, a material thought to be useful as a container of uranium. The hiatus was due to the advisory nature of the Uranium Committee, which had no money of its own and no authority to take action on any of its recommendations. The situation changed dramatically in June 1940, however, when the president, responding to the German invasion of France, formed the National Defense Research Committee (NDRC), under the ongoing leadership of Vannevar Bush, who was given direct access to the president and an independent source of funds. The Uranium Committee was to report directly to Bush.

The NDRC mobilized the scientific resources—the scientists and laboratories—of the nation. At the urging of the army and navy, it excluded foreign-born scientists from formal membership in the Uranium Committee, but it continued to use them as important consultants. The NDRC made arrangements to block publication of reports on uranium research, an idea that originated with the scientists themselves. Most important, the newly strengthened Uranium Committee proposed a plan of research on two key questions that needed immediate answers.

One urgent question was whether a self-sustaining chain of fission reactions could be produced. This would require that at least two fission reactions be generated by each of the neutron products of a single fission reaction, that at least two fission reactions be generated by each of the neutron products of the subsequent fission reactions, and so on. This chain reaction—a nuclear physics term meaning exactly what it says—would be the heart of a self-sustaining nuclear reactor to power either a submarine or an atomic bomb. Multiplication of the number of fissioning nuclei by a factor of at least two in each link of the chain was the necessary condition to release the energy stored in a quantity of uranium by a self-sustaining chain of reactions.

It was not obvious that the multiplication factor of at least two could be achieved even if more than two neutrons were released in every fission. A

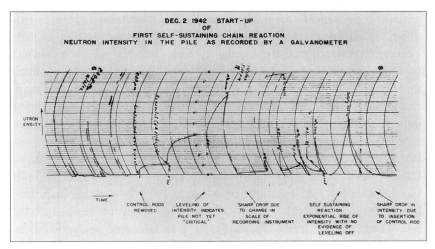

DEC. 2 1942 START-UP
OF
FIRST SELF-SUSTAINING CHAIN REACTION
NEUTRON INTENSITY IN THE PILE AS RECORDED BY A GALVANOMETER

UTRON
ENSITY

TIME

CONTROL RODS
REMOVED

LEVELING OF
INTENSITY INDICATES
PILE NOT YET
"CRITICAL"

SHARP DROP DUE
TO CHANGE IN
SCALE OF
RECORDING INSTRUMENT

SELF SUSTAINING
REACTION
EXPONENTIAL RISE OF
INTENSITY WITH NO
EVIDENCE OF
LEVELING OFF

SHARP DROP IN
INTENSITY DUE
TO INSERTION
OF CONTROL ROD

FIGURE 2.2. *Top*: The S-1 Executive Committee, September 1942. *Left to right*: H. C. Urey, E. O. Lawrence, J. B. Conant, L. J. Briggs, E. O. Murphree, A. H. Compton.

Source: James Phinney Baxter III, *Scientists Against Time* (Boston: Little, Brown, 1946), p. 436.

Bottom: Start-up of the first self-sustaining chain reaction. The figure shows the galvanometer recording of the neutron intensity in the "pile" (reactor) under the stands of the stadium at the University of Chicago on December 2, 1942. Insertion of the control rod after twenty-eight minutes ended the chain reaction.

Source: R. G. Hewlett and O. E. Anderson Jr., *The New World: A History of the U.S. Atomic Energy Commission*, vol. 1, *1939/1946* (Washington, D.C.: U.S. Atomic Energy Commission, 1972), p. 112.

neutron emanating from a fission could do other things besides generating a subsequent fission reaction: it could escape from the uranium-containing vessel, or produce a different nuclear reaction in the uranium, or be absorbed by the material holding the uranium (for example, graphite). Any appreciable loss of neutrons by these means would halt the chain reaction and result only in moderate heating of the uranium sample. Consequently, it was imperative to demonstrate that a self-sustaining chain reaction could be produced in a laboratory.

The second urgent question concerned the physical and chemical properties of uranium and their suitability for bomb production. Most elements occur in nature in the form of isotopes, nuclei with the same chemical properties (same number of protons) but different numbers of neutrons among their constituents. The total number of constituents is, for historical reasons, called the atomic mass of the nucleus and, in conjunction with the ratio of protons to neutrons in the nucleus, determines its stability against disruption by an outside force, for instance, the capture of a neutron from outside. Uranium has two isotopes; one, with an atomic mass of 235, is present in natural uranium at about 0.7 percent of the total, and the other, with an atomic mass of 238, is present in approximately 99.3 percent of the total. Early experiments done with very small quantities of each isotope showed that only the isotope of mass 235 fissioned readily, while the more abundant isotope 238 in the natural element was a limitation and certain to prevent an explosive chain reaction. Consequently, the Uranium Committee's second question addressed the methods of isotope separation that might yield substantial quantities of relatively pure isotope 235. Several methods were known to nuclear scientists at the time, but whether any of these, which had been tested only on minute quantities of material, could be applied to the separation of kilograms of isotope 235 from tons of natural uranium was a completely open question. Large-scale experimentation would be needed.

It was, of course, easier to formulate these questions than to answer them. Organizing the scientists and the laboratories in which the work was to be done, finding the money and supplies that were needed in increasing amounts, and coordinating all of it was more than the combined efforts of the Uranium Committee and the NDRC could accomplish. Both were too small and lacked the power to acquire the necessary resources. In addition, the NDRC had other urgent assignments, in particular, to develop radar and to stimulate research in military medicine. The way out of this impasse was the establishment of the Office of Scientific Research and Development

(OSRD), enacted by executive order on June 28, 1941, as an entity within the Office for Emergency Management of the executive office of the president. Again, Bush was to be its director and was personally responsible to the president. The NDRC would continue, but within the OSRD and with James Conant as its new head. The Uranium Committee became the OSRD Section on Uranium, designated as Section S-1.

There were many reasons to create the OSRD other than the consolidation of the uranium program, but the OSRD acted as the catalyst for the program in all respects. To produce a chain reaction, preparations to build a lattice of graphite and uranium metal and oxide, called a uranium pile, were in progress. Three methods of isotope separation were under test at the engineering level with an eye toward handling large quantities of uranium. The OSRD was also able to encourage study of the newly discovered element plutonium, with an atomic mass of 239: ninety-four protons in its nucleus, two more than in uranium. The significance of this discovery lay in the fact that plutonium was shown to fission under the action of neutrons just as uranium 235 did. Moreover, plutonium 239 and uranium 238, from which it was produced, were different chemical elements and could be separated from each other with far less difficulty than the two almost identical isotopes of uranium. The OSRD and its scientists began to contemplate the prospect of making bombs either of uranium 235 or of plutonium 239 or perhaps some of each kind.

Theoretical and experimental work on all these possibilities went on feverishly throughout the remainder of 1941 and all of 1942. The pressure to produce an atomic bomb before the Germans was immense. On December 2, 1942, to the overwhelming relief of everyone involved, the first controllable, self-sustaining chain reaction was produced in a graphite-uranium pile (a three-dimensional lattice of uranium and graphite rods) located under the stands of the University of Chicago football field. The radioactivity that emerged from the unshielded pile forced the scientists to terminate the chain reaction after a few minutes, but there was no doubting the importance of what had been achieved.

Even before the success in Chicago, the leaders of the Manhattan Project had concluded that construction of an atomic bomb was a feasible undertaking if sufficient funds and manpower were diverted to the effort. In addition to the evidence for a chain reaction, which they anticipated to be forthcoming, they achieved significant progress in uranium isotope separation and in the production and separation of plutonium to support such optimism. Nevertheless, they knew that they were far from manufacturing sufficient quantities of uranium or plutonium for a bomb and had no idea of how to build an actual bomb, much less one capable of a targeted explosion. They understood that a realistic plan would require diversion of resources from the continuing buildup of the U.S. arsenal for the war effort. Furthermore, they were not absolutely sure that a bomb could be produced before the end of the war or even at all, despite their confidence that they were on the right track. The fear that Nazi Germany might successfully do so, however, drove them to press forward and propose to President Roosevelt a tenfold expansion of the atomic bomb project from hundreds of millions to billions of dollars, a monumental sum. Their thinking was summed up in a statement attributed to Ernest Lawrence, a Nobel laureate in physics, head of one of the isotope separation programs, "It will not be a calamity if when we get the answers to the Uranium problem they turn out negative from the military point of view, but if the answers are fantastically positive and we fail to get them first, the results for our country may well be tragic disaster."[4]

The increase proposed in the plan would make the Manhattan Project too big and too costly to remain within the OSRD. Consequently, the plan recommended that an independent organization be formed to administer the project and special funds be provided for it. Complete exchange of atomic bomb information with the British was also recommended. President Roosevelt agreed and formed a new committee, named the Policy Committee, in which all policy relating to the Manhattan Project was to be confined; it consisted of the president; the vice president, Henry Wallace; the secretary of war, Henry Stimson; the army chief of staff, George Marshall; and the two leaders of the OSRD, Vannevar Bush and James Conant, who until then had been responsible for all decisions affecting the Manhattan Project. Rather than creating a new organization, the Policy Committee ordered the army to take control of the atomic bomb project when the uranium and plutonium isotope separation pilot plants—then still under design—were ready to operate, possibly in late 1942. At the same time, following a schedule of unprecedented speed, full-scale plant construction would be started without waiting for the guidance that might be obtained from the pilot plants. In addition, directions were given to construct a new laboratory to be devoted to the task of building an atomic bomb using the fissionable materials furnished by the isotope separation plants. In September 1942 Brigadier General L. R. Groves was placed in charge of all activities relating to the Manhattan Project, designated by the army as the "Manhattan District in the Corps of Engineers." In May 1943 the Manhattan District took over the research and development contracts, as well as the early construction contracts from the OSRD. This marked the end of the organizational connection of the OSRD with the Manhattan Project.

The Manhattan Project enlisted the aid of leading construction companies to build a network of isotope separation plants in isolated areas in sev-

FIGURE 2.3. *Top*: One of the production plants at the Clinton Engineer Works at Oak Ridge, Tennessee.

Source: Henry D. Smyth, *Atomic Energy for Military Purposes: A General Account of the Scientific Research and Technical Development that Went into the Making of Atomic Bombs* (Princeton: Princeton University Press, 1945), plate 7.

Bottom: Leslie R. Groves (left) and J. Robert Oppenheimer.

Source: T. R. Fehner and Jack M. Holl, *Department of Energy, 1977–1994: A Summary History* (Oak Ridge, Tenn.: Office of Scientific and Technical Information, 1995), p. 10.

eral states. These were operated by industries whose peacetime business was large-scale chemical engineering. A full-scale gaseous diffusion plant for uranium isotope separation, operated by the Carbide and Carbon Chemicals Corporation, was built in a Tennessee valley on a 59,000-acre site designated the Clinton Engineer Works, near the Clinch and Tennessee Rivers, eighteen miles west of Knoxville. Two pilot plants, each in a separate valley, one for the production of plutonium and the other using the electromagnetic separation method for uranium, were also built on the same site. The original plan called for a full-scale plutonium plant, but fear that the several piles required for major plutonium production—each very much larger and more powerful than anything built before—might have an accident and endanger the population of Knoxville prompted a change in location to a more isolated area in Washington state. That plant, the Hanford Engineer Works, was located on 670 square miles near the Columbia River at Pasco and operated by the duPont Company. The isolation of these plants, dictated by the need for secrecy and safety, made it necessary to build them from the ground up. This included on-site cities to house as many as sixty thousand construction workers, plant operators, engineers, and scientists.

A laboratory built to learn how to build an atomic bomb—the famous Los Alamos Scientific Laboratory—began to take shape in New Mexico. Again, to preserve secrecy and stay far from populated areas, the laboratory was placed in a remote, relatively inaccessible location and had to be built with extensive housing for families, in addition to buildings for scientific and engineering work.

The nuclear separation plants in Tennessee and Washington began to produce sizable quantities of uranium 235 and plutonium 239 by the end of 1944. Significant progress toward an actual bomb was also made at Los Alamos, but the principal problem there was how to detonate the bomb. If done too slowly—too fast would not be a problem—neutrons would leak away, and the fissionable material would simply fizzle. Ultimate success in producing an exploding bomb seemed assured, but when it would be ready for use was still not very clear. Uncertainty also surrounded the war in Europe and the Pacific. Invasion forces in Europe were engaged in intense fighting in the Battle of the Bulge, and U.S. forces moving north in the Pacific encountered ever-increasing and costly resistance the closer they moved to Japan.

The situation changed dramatically in the spring of 1945. President

Roosevelt died on April 12 and was succeeded by Harry S. Truman. Too late for Roosevelt to witness it, Germany surrendered unconditionally a month later, on May 7. The U.S. Army and Navy were completing plans for their next major operation, the invasion of the island of Kyushu in Japan. At Los Alamos, scientists were in the last stages of preparing to test a uranium bomb at the Alamogordo, New Mexico, Bombing Range, one hundred miles south of Albuquerque. That test, code named "Trinity," went as planned on July 16. It was the first atomic bomb explosion, one with enough destructive power to amaze the scientists and officials who witnessed it.

At the time of the Alamogordo test, President Truman was in Potsdam, Germany, for a meeting with Stalin and Churchill. A major issue before them was how to end the war against Japan. The Soviets had massed an army on the border of Manchuria with the aim of driving the Japanese out of China. The Americans were hoping for Russian assistance against Japan but were unwilling to agree to Stalin's demands for major concessions in both Eastern Europe and China as the price for Russian participation. The success at Trinity changed that. No further persuasion of the Soviets to enter the war against Japan took place. The Potsdam Proclamation of July 26, a last warning to the rulers of Japan, signed by the United States, China, and Great Britain, called on Japan for "the unconditional surrender of all Japanese armed forces [and promised] a peacefully inclined and responsible government [to be established in accord with] the freely expressed will of the Japanese people." No reference was made to the fate of the emperor. The alternative was "prompt and utter destruction." As the conference concluded, Truman casually mentioned privately to Stalin that the United States had a new weapon of unusual destructive force, about which Stalin showed no special interest except to say that he was glad to hear it and hoped the Americans would make "good use of it against the Japanese."[5]

The Japanese government decided to ignore the warning in the Potsdam Proclamation and to "press forward resolutely to carry the war to a successful conclusion."[6] Radio Tokyo began broadcasting this message on July 29, Potsdam time. One week later, on Sunday, August 5, the first atomic bomb was dropped on Hiroshima. The magnitude of the destruction at Hiroshima was initially obscured by the thick layer of dark gray dust that covered the city after the explosion. Because of complete loss of communication with the city, the Japanese government had little idea of the devasta-

FIGURE 2.4. *Top*: Museum display of "Little Boy," the uranium bomb that was dropped above Hiroshima on August 6, 1945.

Source: R. G. Hewlett and O. E. Anderson Jr., *The New World: A History of the U.S. Atomic Energy Commission*, vol. 1, *1939/1946* (Washington, D.C.: U.S. Atomic Energy Commission, 1972), p. 400.

Bottom: Museum display of "Fat Man," the plutonium bomb dropped above Nagasaki on August 9, 1945.

Source: R. G. Hewlett and O. E. Anderson Jr., *The New World: A History of the U.S. Atomic Energy Commission*, vol. 1, *1939/1946* (Washington, D.C.: U.S. Atomic Energy Commission, 1972), p. 400.

tion. It was not until August 7 that the Japanese civilian leaders realized that a single bomb had destroyed the entire city. The Japanese military—believing that little had really occurred—would only agree to send an investigating team to Hiroshima. On August 9, discussions in Tokyo to surrender ended without a decision, despite word that a second atomic bomb attack had destroyed Nagasaki and that the Russians had at last entered the war against them. By morning of the next day, however, amid mass destruction and horror, Japan formally accepted the terms of the Potsdam Proclamation, with the proviso that it would not prejudice the emperor's position. The Japanese military managed to get the U.S. concession that, immediately upon surrender, the authority of the emperor would be subject only to the Supreme Commander of the Allied Powers. Negotiations continued as issues relating to disarmament and occupation were raised once more, but use of the third atomic bomb—the second available plutonium bomb—was expressly forbidden by the president without his consent. On August 10 the Japanese government accepted the Potsdam terms and surrendered.

President Truman briefly summarized for the American people the atomic bomb events in Japan. Ordinarily, he said, the government and the scientists would have made public all technical data, but he did not intend to do so "pending further examination of possible methods of protecting us and the rest of the world from the danger of sudden destruction."[7] There was, however, good reason to issue some kind of technical release if only to keep political pressure and speculation within bounds. A quasi-official report titled *Atomic Energy for Military Purposes* had been prepared earlier by Henry D. Smyth, a professor at Princeton University and a long-time member of the Manhattan Project. Smyth had been asked to write a description of the project from its beginning to its culmination with the aim of informing the public on a matter that, as he saw it, would be of vital concern to them for many years. The report was subtitled "A General Account of the Scientific Research and Technical Development that Went into the Making of Atomic Bombs" but had been carefully scrutinized to make sure it contained no information that might be of value to a foreign nation seeking to produce atomic bombs of its own. Nevertheless, the decision to publish the Smyth report was not easily made. Among the advisers of Secretary of War Stimson, General Groves and James Conant, for example, were in favor of publication because it was a lesser evil; James Chadwick, the British scientific counterpart to Conant, was opposed but

FIGURE 2.5. Henry Stimson, secretary of war during WWII, arrived in Berlin, July 15, 1945. Accompanying him was his aide, Colonel William H. Kyle. Stimson had been secretary of state from 1929 to 1933. He had the confidence of President Roosevelt and Vannevar Bush and was instrumental in the success of the OSRD.

Source: R. G. Hewlett and O. E. Anderson Jr., *The New World: A History of the U.S. Atomic Energy Commission*, vol. 1, *1939/1946* (Washington, D.C.: U.S. Atomic Energy Commission, 1972), p. 392.

pointed out that it was a U.S. matter and anyway would not be of much help to the Russians, and the assistant secretary of war, Robert A. Lovett, was also opposed. The final decision to publish was made by the president on August 12.

The rocky road to publication of the Smyth report was a sign of many fears: fear of the power of the atom if let loose in the world; fear of the absence at the time of a plan for the future of the Manhattan Project and atomic energy in general in the United States; and fear engendered by the emerging threatening nature of the Soviet Union.

FIGURE 2.6. Conference on the Smyth Report. Henry D. Smyth (left) and Ernest O. Lawrence confer at Berkeley, California, autumn 1944.

Source: R. G. Hewlett and O. E. Anderson Jr., *The New World: A History of the U.S. Atomic Energy Commission*, vol. 1, *1939/1946* (Washington, D.C.: U.S. Atomic Energy Commission, 1972), p. 376.

Atomic Energy

for Military Purposes

The Official Report
on the Development of the Atomic Bomb
under the Auspices
of the United States Government,
1940–1945

By HENRY DeWOLF SMYTH

CHAIRMAN, DEPARTMENT OF PHYSICS
PRINCETON UNIVERSITY
CONSULTANT, MANHATTAN DISTRICT, U.S. ENGINEERS

Written at the request of
MAJ. GEN. L. R. GROVES, U.S.A.

PRINCETON

PRINCETON UNIVERSITY PRESS

1945

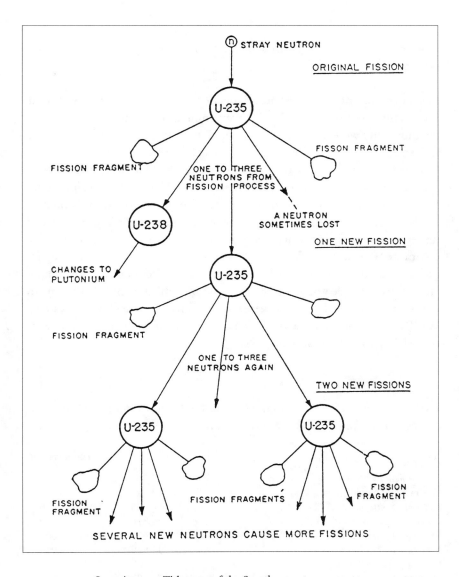

FIGURE 2.7. *Opposite page*: Title page of the Smyth report.

Source: Henry D. Smyth, *Atomic Energy for Military Purposes: A General Account of the Scientific Research and Technical Development that Went into the Making of Atomic Bombs* (Princeton: Princeton University Press, 1945), p. 35.

Above: Schematic diagram of a chain reaction, from the Smyth report.

Source: R. G. Hewlett and O. E. Anderson Jr., *The New World: A History of the U.S. Atomic Energy Commission*, vol. 1, *1939/1946* (Washington, D.C.: U.S. Atomic Energy Commission, 1972), p. 722.

Engineer and educator, Vannevar Bush was the first presidential science adviser.

Many of the events described here happened because of one man: Vannevar Bush. This is not to say that he did everything himself but that his character and ability were instrumental in giving both form and substance to the U.S. science and technology organizations of WWII.

First the NDRC, then the OSRD, and soon after the Manhattan Project bore the stamp of his influence from their earliest days to their termination. While he was director of the OSRD, he was also the senior administrator of the Manhattan Project until, on his own recommendation, it was placed under the direction of the army, although he continued as a member of the committee that set policy at the highest level for the project. In those capacities he served as the principal science adviser to President Roosevelt and, after Roosevelt's death, to President Truman until the end of the war. In effect, he was the architect of organized U.S. science and technology during WWII, just as General George Marshall was the architect of the modernized U.S. Army and its strategy.

Table 2.1. Cumulative Costs in the Manhattan Project Engineer District as of December 31, 1945 (in thousands of dollars)

	PLANT	OPERATIONS
Government overhead	22,567	14,688
Research and development	63,323	6,358
Electromagnetic plant (Y-12)	300,625	177,006
Gaseous-diffusion plant (K-25)	458,316	53,850
Thermal-diffusion plant (S-50)	10,605	5,067
Clinton Laboratories	11,939	14,993
Clinton Engineer Works		
(headquarters and central utilities)	101,193	54,758
Hanford Engineer Works	339,678	50,446
Heavy-water production plants	15,801	10,967
Los Alamos Project	37,176	36,879
Special operating materials	20,810	82,559
TOTALS	1,382,033	507,571

When WWII ended, Bush used his wartime experience to produce a blueprint, *Science: The Endless Frontier*, for organizing science and technology in the United States in peacetime, just as General Marshall produced the peacetime plan that bore his name to organize the recovery of Europe from the destruction of the war.

Long after WWII, Jerome B. Wiesner, who had been science adviser to presidents Kennedy and Johnson, began a biographical memoir of Bush with these paragraphs.

No American has had greater influence in the growth of science and technology than Vannevar Bush, and the twentieth century may yet not produce his equal. He was an ingenious engineer and an imaginative educator, but above all he was a statesman of integrity and creative ability. He organized and led history's greatest research program during World War II and, with a profound understanding of implications for the future, charted the course of national policy during the years that followed.

The grandson of two sea captains, "Van" Bush manifested his Cape Cod heritage in a salty, independent, forthright personality. He was a man of strong opinions, which he expressed and applied with vigor, yet he stood in awe of the mysteries of nature, had a warm tolerance for human frailty, and was open-minded to change and to new solutions to problems. He was pragmatic, yet had the imagination and sensitivity of a poet, and was steadily optimistic.[8]

Wiesner had a long, intimate association with MIT, where Bush began his career as an administrator of science and technology. Wiesner joined an OSRD laboratory, the MIT Radiation Laboratory, in 1942; almost thirty years later he became president of MIT. He was a friend and colleague of Vannevar Bush, with a close-up view of the man and his accomplishments. At the same time, he developed a perspective on American science from his experience as a presidential science adviser.

From Bush's own writing about how the OSRD and Manhattan Project were created and functioned, he emerges as a man with extraordinary good sense and superb organizational ability. He was able to develop and maintain good working relations with government agencies and Congress and to win and keep the respect and support of President Roosevelt and of the exceptionally influential secretary of war, Henry L. Stimson. And he retained the confidence of the leaders of the scientific community, themselves eminent spokesmen for science and engineering. Perhaps his distinctive features were a lack of pretension, selfless dedication to the cause at hand, and a good sense of humor. He included in his 1970 book *Pieces of the Action* a chapter on the critical need for proper organization to accomplish complex tasks. It begins, "When Eve joined Adam there was formed the first organization in history. It was a simple one, yet its essential relations and the regulations governing it have not even today been fully worked out."[9] It is an intriguing and effective way to begin a discussion of what could be a relatively dry subject.

Vannevar Bush grew up near Boston and graduated from Tufts College. He received a doctorate in engineering from a joint Harvard-MIT program in 1916. When the United States entered WWI in 1917, Bush worked on methods of submarine detection until the end of the war. He saw first hand how narrow the Allied victory over the submarine had been, and he learned about human relations problems in a research and development effort that involved both civilian scientists and military personnel.

Three years later, Bush joined the engineering faculty of MIT, where, in addition to teaching, he served as a consultant to a small firm from whose patents grew the successful electronics corporation Raytheon. At that time he also turned to the invention of electromechanical calculating machines that simulated complicated problems in physics and engineering, allowing quantitative solutions to be obtained. An important, relatively new problem of his time was how to optimize the efficiency of a network consisting of electrical power–generating stations, transmission lines, and the electric power loads on them. For this work, accomplished with a team of graduate students, Bush won a medal from the Franklin Institute of Philadelphia in 1928. He maintained a strong interest in calculating machines throughout his life. In the summer of 1945, as WWII was coming to an end, he published an article that described a machine called the "memex" that was remarkable as a forerunner in concept and detail of the ubiquitous personal computer of a half century later.

Bush was named dean and vice president of engineering of MIT in 1932 and might have become its president, but in 1938, foreseeing the coming war and U.S. involvement in it, he left to become president of the Carnegie Institution of Washington. This move opened the highest levels of national science policy making to him. Soon afterward, he was appointed to the

National Advisory Committee for Aeronautics (NACA) and a year later became its chairman. All these posts polished Bush's natural talent as an administrator of science and an adviser on science policy.

The experience with NACA fixed in Bush's mind the need for the chairman of a federal science advisory committee to report directly to the president and to receive support through emergency funds available to the president or by congressional legislation. The alternative would be an advisory committee without a strong sponsor and without federal funding, in short, without the capability to effect real change.

In June 1940 President Roosevelt established the National Defense Research Committee (NDRC) to coordinate the nation's science resources and supplement the weapons development programs of the army and navy. This was done in response to recommendations made by Bush, who was appointed chairman of the new committee. In this capacity he was the agent among many groups poised to face the challenges of wartime: the science community, the White House, Congress, and the armed services. The NDRC was supported by the president's emergency funds, and Bush reported directly to the president. In effect, Bush was the president's science adviser, the first of a long line of individuals to serve in that capacity in future administrations.

One year later, again in response to recommendations by Bush, the NDRC was incorporated in the newly created OSRD, which included the Uranium Committee then charged with evaluating the prospects for an atomic bomb and a section on medical research. The OSRD had the authority to promote the manufacture of the devices that emerged from its research, as it did in early nuclear fission experiments and with blood derivatives. The creation of the OSRD was a measure of Bush's vision and of his success in dealing directly with the president, managing a complex budget, and working with the Congress. The accomplishments of the OSRD were a testimonial to Bush's talent for bringing together civilian and military leaders in diverse areas of science, providing them with the organization and funds to enable them to work productively on joint projects, and stimulating them to do so harmoniously.

Bush tells a story in *Pieces of the Action* that reflects his view of what he did during the war. Some time after the National Science Foundation (NSF) had been established in 1950, there was a dinner at which Bush presided and President Truman spoke. They sat together at dinner, and in the course of their conversation the president asked Bush to serve as a member of the NSF

FIGURE 2.8. Roosevelt and Churchill at Quebec. The man in uniform is the earl of Athlone, then governor-general of Canada. In the background is Canadian prime minister Mackenzie King.

Source: R. G. Hewlett and O. E. Anderson Jr., *The New World: A History of the U.S. Atomic Energy Commission*, vol. 1, *1939/1946* (Washington, D.C.: U.S. Atomic Energy Commission, 1972), p. 272.

Science Board, which was then being formed. Bush demurred, probably because he and Truman did not mesh personally as he had with Roosevelt and because of their disagreement over the method of selection and duties of the members of the Science Board. Truman finally agreed to leave him off the board. Then he said, "Well, Van, you are not looking for a job, are you?" And Bush replied, "No, Mr. President, I am not looking for a job." The president added, "You cannot say I went looking for this job that I am in," and Bush commented, "No, Mr. President, not the first time," which obliquely referred to the 1948 presidential campaign in which, contrary to all predictions, Truman, the incumbent, defeated Dewey, the challenger. The president was tickled and, poking Bush in the ribs, said, "Van, you should be a politician. You have some of the instincts." To which Bush responded, "Mr. President, what the hell do you think I was doing around this town for five or six years?"[10]

Bush finished his report *Science: The Endless Frontier* in 1945. It had a profound impact on Washington politics even though the president had not requested it. Few disagreed with the need for a federal science agency or with the method proposed to provide one. The political climate of Washington was rapidly returning to the peacetime mode, however, and approval of an idea in principle no longer led to its quick realization in practice. Furthermore, Bush failed to recognize how seriously Truman regarded presidential control of the director and members of the governing board of any federal science agency. In fact the president vetoed a bill to establish a national science foundation that emerged from Congress in 1947 because of disagreement on that very issue. Three more years would pass before the foundation became a reality.

Bush was active in Washington until the foundation was established and for some time thereafter. True to his word, he "was not looking for a job," and in 1955 he resigned his position as head of the Carnegie Institution and returned to his home in Massachusetts. Nevertheless, he was regularly called on for testimony before congressional committees and for advice and recommendations as the postwar science agencies grew.

His intellectual drive did not abate. In the period between 1952 and 1959 he published articles on science in medicine, on an electric micromanipulator, on an automatic microtome, and on the surgical correction of calcification of the aorta in adults. And, of course, his interest in calculating machines remained. In addition, he published on the organization and administration of military research programs, on improving the

SCIENCE

THE ENDLESS FRONTIER

A Report to the President

by

*V*ANNEVAR *B*USH

Director of the
Office of Scientific Research and Development

•

July 1945

United States Government Printing Office
Washington : 1945

FIGURE 2.9. Title page of the Bush report, *Science: The Endless Frontier*.

Source: Vannevar Bush, *Science: The Endless Frontier* (Washington, D.C.: U.S. Government Printing Office, 1945).

patent system, and on the relation of fundamental research to engineering. He wrote three books: *Science Is Not Enough*, *Modern Arms and Free Men*, and *Pieces of the Action*. In the last, he reminisced about his experiences in and out of the government. Bush died in June 1974, at age eighty-four.

FIGURE 2.10.   Vannevar Bush, director of the Office of Scientific Research and Development and first presidential science adviser.

Source: Richard Mandel, *A Half Century of Peer Review (1946–1996)* (Alexandria, Va.: Division of Research Grants, National Institutes of Health, Logistic Applications, 1996), p. 8.

CHAPTER 3

Courtship: 1945–1955

The end of the war brought joy, relief, and new challenges to the United States. The nation's wartime accomplishments were harbingers of a bright future, and its wealth was at last free to be invested in that future. But the challenges were many, and the ways to meet them unclear. The United States was in a position similar to that of the winner of a very big lottery prize. A new way of life lay within its grasp if—and it was a big if—the unfamiliar new wealth could be safeguarded and invested wisely.

The challenges in science and technology were especially pointed because the close connection between atomic bombs and national security was obvious, and the impact of wartime advances in medicine was already part of U.S. daily life. It was natural to ask what the government could do to preserve and expand the institutions that produced these and other benefits. This question, so much easier to ask than to answer, would occupy Washington for many years after the war. The answer would come as a series of seemingly disconnected actions. These began in 1945 with the report from Bush to President Roosevelt proposing a federally funded foundation dedicated to the support of science in U.S. universities. Washington, however, was occupied with science problems that had carried over from the war because they were directly related to national security. Solutions to those problems involved creation of several science agencies ostensibly different from Bush's proposed foundation that nevertheless functioned in the manner he had foreseen. In this way, almost inadvertently, scientists and the government began to establish a close relationship during the five-year period before the passage of the actual act that created the National Science Foundation.

In the beginning, there was the Bush report: Science: The Endless Frontier

One way or another, a majority of scientists in the United States had been involved in the OSRD or the Manhattan Project during WWII. They experienced the power that science and technology could wield when self-organized and directed toward definite goals, in each instance to produce an object or device to do a specific job. They were well aware, however, that underlying the accomplishments of applied research during wartime was the scientific knowledge acquired from basic research, done mostly for its own sake in the years before WWII. Many who worked to develop radar, for example, knew of experiments that studied the ionosphere—a layer of electrically charged atoms surrounding the earth—by reflecting radio waves from it. These experiments, performed as early as 1925, were motivated by interest in the properties of the earth's atmosphere, not by the idea that they might lead to a method for tracking aircraft more than a decade later. In like fashion, the chemists who did the basic research on the properties of proteins during the twenty-year period before WWII pursued their work because proteins were recognized to be fundamental biochemical compounds, not because they would provide blood and blood derivatives for battlefield transfusions decades later.

Similarly, the achievement of the Manhattan Project was rooted in more than ten years of study of the atomic nucleus, a subscience of physics, pursued for its own sake and thought then to be remote from any practical application. There is a story of the effect of a lecture given at a meeting of the American Physical Society in 1939 by Niels Bohr, the world-famous Danish physicist. Bohr had become aware of the experiments in Germany that demonstrated the fission of uranium, and he readily understood their significance. He described the experiments to the U.S. audience. Before he finished his lecture, so the story goes, physicists in the audience began to rush out of the hall either to telephone their laboratories or to return directly to them to initiate experiments that would test the validity of the information Bohr had just given them. These experiments, which effectively launched the United States into the Manhattan Project, could not have been done so quickly and conclusively without the years of basic research that preceded them. Nor could the work of the Manhattan Project have prospered without that fund of knowledge.

Of course, the work of both the OSRD and the Manhattan Project also involved the acquisition of substantial new knowledge, which illustrates the difficulty of making a clear-cut distinction between applied and basic research, particularly in cases where they merge almost seamlessly into one another. Nevertheless, most scientists agreed that the research done during WWII was for the immediate purpose of reaching practical goals, and in that respect it was applied research. They also agreed that its success rested on the strong foundation of basic research done without practical goals in mind. No one was more aware of this than Vannevar Bush and his colleagues in the OSRD and the Manhattan Project. They recognized that the successful technology of wartime—indeed of any time—depended on access to a flourishing national resource of basic scientific research. This in turn led them to the belief that the federal government had to provide support for basic research in peacetime as it had for applied research in wartime. Although basic research is not easily justified by short-term practical accomplishments, they had become convinced that it was an essential component of modern technological progress and therefore vital to the health and security of the nation in peacetime. The first step taken to bring the government into peacetime science was the Bush report, *Science: The Endless Frontier*, written in response to the following letter from President Roosevelt.

The White House

November 17, 1944

Dear Dr. Bush:

The Office of Scientific Research and Development, of which you are the Director, represents a unique experiment of team-work and cooperation in coordinating scientific research and in applying existing scientific knowledge to the solution of the technical problems paramount in war. Its work has been conducted in the utmost secrecy and carried on without public recognition of any kind: but its tangi-

ble results can be found in the communiques coming in from the bat-
tlefronts all over the world. Some day the full story of its achievements
can be told.

There is, however, no reason why the lessons to be found in this
experiment cannot be profitably employed in times of peace. The
information, the techniques, and the research experience developed
by the Office of Scientific Research and Development and by the
thousands of scientists in the universities and in private industry,
should be used in the days of peace ahead for the improvement of the
national health, the creation of new enterprises bringing new jobs,
and the betterment of the national standard of living.

It is with that objective in mind that I would like to have your recommendations on the following four major points:

First: What can be done, consistent with military security, and
with the prior approval of the military authorities, to make known
to the world as soon as possible the contributions which have been
made during our war effort to scientific knowledge?

The diffusion of such knowledge should help us stimulate new
enterprises, provide jobs for our returning service men and other
workers, and make possible great strides for the improvement of the
national well-being.

Second: With particular reference to the war of science against
disease, what can be done now to organize a program for continuing in the future the work which has been done in medicine and
related sciences?

The fact that the annual deaths in this country from one or two
diseases alone are in excess of the total number of lives lost by us in
battle during this war should make us conscious of the duty we owe
future generations.

Third: What can the Government do now and in the future to
aid research activities by public and private organizations: The
proper roles of public and of private research, and their interrela-
tion, should be carefully considered.

Fourth: Can an effective program be proposed for discovering
and developing scientific talent in American youth so that the
continuing future of scientific research in this country may be
assured on a level comparable to what has been done during the
war?

New frontiers of the mind are before us, and if they are pioneered with the same vision, boldness, and drive with which we have waged this war we can create a fuller and more fruitful employment and a fuller and more fruitful life. I hope that, after such consultation as you may deem advisable with your associates and others, you can let me have your considered judgment on these matters as soon as conven-ient—reporting on each when you are ready, rather than waiting for completion of your studies in all.

Very sincerely yours,
Franklin D. Roosevelt

Bush's report and the letter of transmittal that went with it were not delivered to President Roosevelt. Between the president's letter of November 1944 and Bush's report of July 1945, Roosevelt had died of exhaustion and a massive cerebral hemorrhage, on April 12, 1945. Harry S. Truman, essentially a stranger to Bush, assumed the presidency, and his relationship with Bush would prove to be very different from the one Bush had with Roosevelt. Bush's letter, which follows here, outlined his personal frame of reference and the method he selected to prepare the report. He took full responsibil-ity for its recommendations and the mechanics of implementing them.

July 5, 1945

Dear Mr. President:

In a letter dated November 17, 1944, President Roosevelt requested my recommendation on the following points:

(1) What can be done, consistent with military security, and with the prior approval of the military authorities, to make known to the world as soon as possible the contributions which have been made during our war effort to scientific knowledge?

(2) With particular reference to the war of science against disease, what can be done now to organize a program for continuing in the future the work which has been done in medicine and related sciences?

(3) What can the Government do now and in the future to aid research activities by public and private organizations?

(4) Can an effective program be proposed for discovering and developing scientific talent in American youth so that the con-tinuing future of scientific research in this country may be

assured on a level comparable to what has been done during the war?

It is clear from President Roosevelt's letter that in speaking of science he had in mind the natural sciences, including biology and medicine, and I have so interpreted his questions. Progress in other fields, such as the social sciences and the humanities, is likewise important; but the program for science presented in my report warrants immediate attention.

In seeking answers to President Roosevelt's questions I have had the assistance of distinguished committees specially qualified to advise in respect to these subjects. The committees have given these matters the serious attention they deserve; indeed, they have regarded this as an opportunity to participate in shaping the policy of the country with respect to scientific research. They have had many meetings and submitted formal reports. I have been in close touch with the work of the committees and with their members throughout. I have examined all of the data they assembled and the suggestions they submitted on the points raised in President Roosevelt's letter.

Although the report which I submit herewith is my own, the facts, conclusions, and recommendations are based on the findings of the committees which have studied these questions. Since my report is necessarily brief, I am including as appendices the full reports of the committees.

A single mechanism for implementing the recommendations of the several committees is essential. In proposing such a mechanism I have departed somewhat from the specific recommendations of the committees, but I have since been assured that the plan I am proposing is fully acceptable to the committee members.

The pioneer spirit is still vigorous within this nation. Science offers a largely unexplored hinterland for the pioneer who has the tools for his task. The rewards of such exploration both for the Nation and the individual are great. Scientific progress is one essential key to our security as a nation, to our better health, to more jobs, to a higher standard of living, and to our cultural progress.

Respectfully yours,
V. Bush, Director

Science: The Endless Frontier *discussed the purpose of a federal science agency and presented specific recommendations for its organization and functions.*

It consisted of six parts whose titles reveal the pattern of Bush's thinking and whose contents were specific enough to provide plans of action: "Introduction," "The War Against Disease," "Science and the Public Welfare," "Renewal of Our Scientific Talent," "A Problem of Scientific Reconversion," and "The Means to the End."

The major recommendation of the report was made in part 6, "The Means to the End": "The federal Government should accept new responsibilities for promoting the creation of new scientific knowledge and the development of scientific talent in our youth." Bush went on to say, "The effective discharge of these responsibilities will require the full attention of some over-all agency devoted to that purpose, and there should be a central point within the Government for a concerted program of assisting scientific research conducted outside of Government." Bush emphasized that "the agency should furnish funds needed to support basic research in the colleges and universities; should coordinate where possible research programs on matters of utmost importance to the national welfare; and should formulate a national policy for the Government toward science." This was his proposal for a national research foundation to be established by Congress.

The membership, functions, and initial level of funding of the foundation were outlined in detail. Responsibility for the foundation would be in the hands of nine "Members of the Foundation" who would not otherwise be connected with the government or represent any special interest. They were to be selected by the president to promote the purposes of the foundation. The members were to serve four-year terms, choose their chairperson annually, and be reimbursed for their expenses only. The responsibilities of the members and of the chief executive officer, the director of the foundation, who was to be selected and appointed by the members, were also specified. The foundation would consist of several professional divisions responsible to the members of the foundation: medical research, natural sciences, national defense, scientific personnel and education, and publications and scientific collaboration. The functions and authority of the divisions were spelled out at length, so that the proposed foundation was more than a skeleton organization. The proposal could be considered as a draft of legislation for Congress.

Bush was careful to note the foundation's need for special authority. He argued that the foundation should be free from the obligation to put its contracts for research out for bids, since the measure of a successful research contract should not be its dollar cost but its contribution to knowledge. He also asserted that the foundation should be free to place its research contracts and grants with institutions whose latent talent or creative atmosphere would afford the promise of research success, as well as with institutions that had demonstrated research capability. As in wartime research sponsored by the OSRD, he proposed that the research sponsored by the foundation be conducted on an actual cost basis, including appropriate overhead, but not for profit.

Bush provided a table of rough estimates of the budgets for the foundation's first and fifth years, after which he expected operation would reach a stable level. In 1945 his dollar amounts were realistic, although a $50 million budget for the Division of Natural Sciences in its fifth year shocked many in Congress. Later, these costs would reflect the effects of progress and inflation.

Bush argued that the creation of the foundation was so important that prompt action by Congress was necessary. Legislation drafted with great care and speed was imperative if the nation was to meet the challenge of science and fully utilize its potential without losing momentum in the

Table 3.1 Budget Estimates for the NRF for the First and Fifth Years

	ACTIVITY FIRST YEAR	MILLIONS OF DOLLARS FIFTH YEAR
Division of Medical Research	5.0	20.0
Division of Natural Sciences	10.0	50.0
Division of National Defense	10.0	20.0
Division of Scientific Personnel and Education	7.0	29.0
Division of Publications and Scientific Collaboration	0.5	1.0
Administration	1.0	2.5
TOTALS	33.5	122.5

Source: Vannevar Bush, Science: The Endless Frontier (Washington, D.C.: U.S. Government Printing Office, 1945), p. 33.

transition from war to peace. Since its organization was not without prece-dent—it was patterned after the successful National Advisory Committee for Aeronautics (NACA), which had promoted basic research in the prob-lems of flight during the previous thirty years and with which Congress was familiar—such legislation, he averred, should be possible.

If the recommendation to create a national research foundation was the heart of the report, then its soul was contained in the section titled "Five Fundamentals." Indeed, in his two-paragraph request for action by the Congress, Bush made it dramatically clear that "whatever program [for the foundation] is established it is vitally important that it satisfy the Five Fun-damentals":

> There are certain basic principles which must underlie the program of Government support for scientific research and education if such sup-port is to be effective and if it is to avoid impairing the very things we seek to foster. These principles are as follows:
>
> (1) Whatever the extent of support may be, there must be stability of funds over a period of years so that long-range programs may be undertaken.
> (2) The agency to administer such funds should be composed of citizens selected only on the basis of their interest in and capacity to promote the work of the agency. They should be persons of broad interest in and understanding of the peculiarities of scientific research and edu-cation.
> (3) The agency should promote research through contracts or grants to organizations outside the federal Government. It should not operate any laboratories of its own.
> (4) Support of basic research in the public and private colleges, universi-ties, and research institutes must leave the internal control of policy, personnel, and the method and scope of the research to the institu-tions themselves. This is of the utmost importance.
> (5) While assuring complete independence and freedom for the nature, scope, and methodology of research carried on in the institutions receiving public funds, and while retaining discretion in the alloca-tion of funds among such institutions, the Foundation proposed herein must be responsible to the President and the Congress. Only through such responsibility can we maintain the proper relationship between science and other aspects of a democratic system. The usual controls of audits, reports, budgeting, and the like, should, of course,

apply to the administrative and fiscal operations of the Foundation, subject, however, to such adjustments in procedure as necessary to meet the special requirements of research.

Basic research is a long-term process—it ceases to be basic if immediate results are expected on short-term support. Methods should therefore be found which will permit the agency to make commitments of funds from current appropriations for programs of five years duration or longer. Continuity and stability of the program and its support may be expected (a) from the growing realization by the Congress of the benefits to the public from scientific research, and (b) from the conviction which will grow among those who conduct research under the auspices of the agency that good quality work will be followed by continuing support.

Two of the recommendations in particular were strikingly new. The first was the notion that the principal function of the new agency should be to furnish funds to support basic research outside of the government. That idea was emphasized throughout the report. For example, in part three, "Science and the Public Welfare," Bush argued in the subsection "Science and Jobs" that the nation would not reach full employment or increase the production of goods and services by standing still, by making the same things made before and selling them at the same or higher prices. The solution, argued Bush, lay in basic research performed without thought of practical ends, which would lead to general knowledge and an understanding of nature and its laws. This would provide the means of answering important practical problems without which the progress of industrial development would eventually stagnate. And Bush was careful to hammer home the points that a peculiarity of basic science has always been the variety of paths that lead to new knowledge and that many of the most important discoveries had come as a result of experiments undertaken with very different purposes in mind.

Second, the report insisted that colleges, universities, and endowed research institutes would be the principal institutions to furnish new scientific knowledge and trained research workers. The reasons for this choice were simple and straightforward: "They are uniquely qualified by tradition and by their special characteristics to carry on basic research; . . . scientists may work in an atmosphere which is relatively free from the adverse pressure of convention, prejudice, or commercial necessity." Bush knew that scientists in those institutions were led to study phenomena by curiosity and instinct that guided them to experiments and observations that occasionally opened vast areas of entirely new information. Within three years of his report, as if designed to validate his claim, university scientists conducted several deeply revealing experiments. One of these involved renewed, more accurate observation of the invisible, high-energy cosmic ray particles that constantly bombard Earth, among which were found elementary particles that had never before been observed. These results, products of basic research with no foreseeable practical application, brought recognition of an entirely new particle universe and began the new subscience of elementary particle physics.

In contrast, the report went on to say, scientists employed in research laboratories that were supported by individual industries—with the aim of improving their products and developing new ones—were usually inhibited from pursuing wide-ranging basic research. Industrial laboratories generally have clearly defined standards and goals and are subject to the constant pressure of commercial necessity, all of which limit the opportunity for scientists to engage in basic science. Furthermore, the report found that the government laboratories of 1945 were not qualified for basic research because most of their research was applied, that is, directed toward results that were important to the operations of the government agencies that supported them and not to the increase of basic, general knowledge.

Although Bush was firmly convinced that basic science research and education would be done best in universities, he knew firsthand the value of applied research in peacetime. He wanted to make sure that industries with laboratories focused on applied research were encouraged to continue and even extend their support of those laboratories. From his own experience, Bush knew the increase in the cost of doing business when a research laboratory was supported by an industry. It was an expensive proposition and, without some compensatory help by the government, likely to be cut back or possibly terminated in difficult financial times. He was also aware that industry, when allocating resources for research, was often dependent on the extent to which the government provided financial help. No one proposed that direct government support be given, but Bush suggested that the Internal Revenue Tax Code should be amended to provide for the deduction of expenditures for research and development as current charges against net income. This would serve as a significant incentive.

Bush directly addressed the question of government research laboratories as he had the question of industrial laboratories. They would not be sources of basic research; nevertheless, government agencies and departments faced practical, technical problems in peacetime that required solutions only systematic research could provide. Many in Washington recognized this after the war. Bush understood that government research directed toward well-stated, well-justified goals would prosper under the watchful eye of Congress without any special pleading.

The government traditionally regarded research and education as two avenues of undesirable entry into the private lives and values of its citizens. Time-honored American insistence on the separation of church and state carried over into the minds of many to a separation of school and state. There was long-standing fear that subsidies by the government would lead to domination by the government and ultimately to a single state religion or a single state educational system. Bush felt that the accomplishments of science and technology in WWII and their promise for peacetime presented a unique opportunity to deal directly with that fear. He realized that his most persuasive argument consisted of the proposed agency that would fund scientific research in universities under the auspices of both the executive and legislative branches of the government, either of which could protect against excessive zeal or misbehavior. *Science: The Endless Frontier* outlined his proposal to do so. It was a distillation of the important lessons of WWII that needed to be taken to heart in peacetime. Those lessons had also been learned by others in the government.

Before any federal science foundation could be created, however, Congress, the Truman administration, and scientists had to find common ground on which to build it. A strength of Bush's report was the detailed nature of the plan it presented, but that was also a weakness because the devil lay in the details. It seemed that everyone in Washington had suggestions for change, some major, some minor, that needed to be aired and debated. But the real reason for a delay of five years before Congress passed and Truman signed the National Science Foundation Act was not quibbles over details. Creation of any science foundation was intimately tied to the fates of the Manhattan Project and the OSRD. Congress knew that it had to attend to the future of the Manhattan Project before it did anything else concerned with the future of science and technology in the United States, and Bush knew it too.

The Atomic Energy Commission and the Joint Committee on Atomic Energy of the House and Senate were created to take control of atomic energy and its future.

Informal planning for the postwar future of the Manhattan Project began well before the end of the war, but the call for legislative action after the surrender of Japan found Congress not yet ready to act. The issues were too numerous and broad, and Congress needed an extended period of preparation to come to grips with them. One difficulty was how to define the limits within which scientific information only—not weapons information— might be exchanged with other nations. A clear statement would make possible a proposal from the president to the United Nations "under which cooperation might replace rivalry in the field of atomic power." That position had been advocated by Stimson before he left office and was taken up by the president and Secretary of State James Byrnes. Some language to that effect was likely to be in any bill that would receive serious consideration, but the scope of the offer to share U.S. knowledge was a matter of dispute.

The first bill off the mark was sponsored by Edwin C. Johnson, the ranking member of the Senate Military Affairs Committee and Andrew J. May, chairman of the corresponding committee in the House. Their bill had been prepared within the War Department at the direction of the secretary of war, Robert P. Patterson, who had succeeded Stimson, but the bill had supporters from a variety of backgrounds including some of the leading wartime science administrators. The proponents of the May-Johnson bill hoped for quick action in the Senate, but Senator Arthur Vandenberg objected to assigning the bill to the Military Affairs Committee on the ground that properly it should be considered by a special joint committee of Senate and House competent on the issues. This stalled the bill in the Senate. The House moved rapidly, however, since May was chairman of his committee, and hearings began as soon as the bill was submitted. Testifying on the first day were Patterson, General Leslie Groves, Bush, and James Conant, after which the committee, to everyone's surprise, went into executive session to consider the bill on the same day.

The testimony and questions from the committee had concentrated on the maintenance of security and on the broad powers of control of every aspect of atomic energy. This emphasis and the brevity of the hearing alienated many scientists in Manhattan Project plants and laboratories. They were leery of the strong influence of the army in the May-Johnson bill and believed that Bush and Conant were misguided in supporting it. These objections brought about a resumption of the hearings, and physicists Leo Szilard, A. H. Compton, and J. R. Oppenheimer gave testimony. Szilard was the only scientist to testify before the May committee during the two days of hearings who had not held an administrative position in the Manhattan Project. He was an individual of considerable originality and insight, a seer into the future, and a clever scientist. One of the first to recognize the possibility of a fission bomb, he had been instrumental in persuading Einstein to write his famous letter to Roosevelt. His advice was sought, but he was neither strongly for nor against the House bill; it was not an arena in which his brilliance would shine. For his part, Compton agreed that control was important but preferred to stress development of atomic energy more than the bill did. Still, he felt that the bill was satisfactory. Oppenheimer, on the other hand, suggested that the bill should define the powers of the future Atomic Energy Commission (AEC) more sharply but observed that he could support it because, as he put it, it would get the army out of atomic energy.

Nevertheless, the May-Johnson bill gathered opponents as time went on. Scientists, public figures, and citizen groups urged its withdrawal to provide an opportunity for more public debate on this critical issue. Mistrust of the main features of the bill—emphasis on secrecy and control and loosely specified limits on the powers of the proposed commission—tended to put people off, as did the effort to rush the bill to passage. Meanwhile, the Senate created an eleven-man select committee of its own with a freshman senator, J. Brien McMahon, as chairman a few days after the House committee hearings began. The White House staff took a second look at the May-Johnson bill, specifically at the limits it placed on the president's authority to control the proposed commission. At earlier hearings, the claim had been made that the bill represented the views of the administration, but the Bureau of the Budget and the Office of War Mobilization and Reconversion (OWMR), speaking for the White House, questioned that claim. They argued that the bill would make the commission essentially independent of executive control, because commission members would serve nine-year terms and only three positions would come up for appointment in any administration. Possibly more important, the president's power to remove a commissioner, even for good cause, was limited. Truman reacted to this information by qualifying the White House commitment to the bill.

By this time, support for the bill even within the May committee had become frayed. Amendments were proposed to have the commission head and other commissioners appointed by the president and for each to serve an indeterminate term at the president's pleasure. To encourage research and development in atomic energy, the bill's restrictions on nonweapon research outside the commission were loosened. These amendments gave rise to a majority report and two minority reports from the committee.

Here was a good opportunity for the select committee of the Senate to enter the debate. However, the chairman, McMahon, first had to overcome the committee's lack of knowledge of the elements of atomic energy and the broader issues involved. He did this by soliciting advice from scientists, choosing as consultant to the committee a veteran of the Manhattan Project, Edward U. Condon, recently appointed director of the National Bureau of Standards, who with others conducted a series of tutorial sessions on atomic energy for the select committee. McMahon found an important ally in James R. Newman, a young lawyer in the OWMR, who began work on the draft of a bill to replace the May-Johnson bill. By the end of 1945, that draft, McMahon's bill, was ready to be released to the public.

Learning from objections to the May-Johnson bill, McMahon and Newman proposed an exclusively civilian commission of five full-time members appointed by the president with consent of the Senate, serving indefinite terms at the pleasure of the president. Four mandated divisions of the commission—research, production, materials, and military applications—would each have a director appointed by the president to ensure adequate attention by the White House.

The bill insisted that production and stockpiling of fissionable material remain strictly within control of the commission, which would, however, be allowed to finance basic research by nongovernment institutions in the physical, biological, and social sciences. There would be a minimum of restriction on the flow of information. Basic scientific information would be completely in the public domain, while related technical information would be published to the extent consistent with national security. The commission would hold all patents relating to the production of fissionable material and weapons, but patents covering devices or processes utilizing atomic energy would be subject to compulsory, nonexclusive licensing to prevent private or government monopoly.

For the most part, these provisions were received favorably in the country. Scientists were pleased by the emphasis on civilian control and the freedom to finance and conduct research and disseminate scientific information. Bush and Conant had spoken in favor of the May-Johnson bill because it limited the commission's freedom to carry out or finance research; they believed that research and the business of operating plants and building weapons were largely incompatible functions. Research, they argued, should be sponsored exclusively by an agency dedicated to that purpose, like the National Science Foundation, also under consideration in the Senate at that time. On the other hand, the production of fissionable materials and military applications should be the full-time province of the commission. Many scientists and other professionals, however, were frightened by a commission closed off from the scientific community but open to the military. Public interest in atomic energy and a bill to regulate it was heightened by the activities of newly formed organizations of scientists, particularly the Federation of Atomic Scientists, which rallied women's groups, labor unions, and religious and civic organizations to speak out in favor of the mixture of civilian administration and scientific freedom that scientists and educators found in the McMahon bill.

By February 1946 both the May-Johnson and McMahon bills had gone through a series of modifications, and each appeared in Congress in near-final form. They were compared in public debate by Henry Wallace, Truman's secretary of commerce and Roosevelt's former vice president. Wallace had been a member of Roosevelt's Policy Committee for the Manhattan Project and was in a position to speak with authority on most aspects of atomic energy, particularly the economic implications for peacetime. He testified before the select committee in support of the McMahon bill, insisting that the ultimate international control of atomic energy provided for in the bill was the only alternative to an atomic arms race. For this reason he supported civilian control of the commission, free exchange of basic scientific information, and early development of methods for international inspection of atomic energy activities. Wallace approved of the bill's recommendation to foster the peaceful uses of atomic energy and its patent provisions.

In contrast, Wallace found the May-Johnson bill to be inconsistent with the administration's avowed policy of eventual international control of atomic energy. The bill was intended to promote military development of atomic energy. It would place sweeping powers in the only full-time administrator of the commission (all others being part-time), who might be a military officer, subject to removal by the president only with difficulty. Finally, Wallace said, the May-Johnson bill placed far too little emphasis on the development of peaceful uses of atomic energy.

The other voice from the president's cabinet was that of Secretary of War Patterson. He stressed the Pentagon's objections to the McMahon bill's investment of exclusive control of production and ownership of all fissionable materials in the commission. This, Patterson stated, would inevitably lead to serious complication and confusion in the use of atomic weapons in an emergency. The Pentagon was also concerned by the failure of the bill to provide for a general manager of the commission to ensure efficiency of its operations and by the inadequate penalties for careless or intentional mishandling of weapons information.

Patterson was strongly opposed to the exclusion of the armed forces from the commission since it would leave the military with the responsibility for delivering atomic weapons but without authority to produce or control them. This issue of military exclusion was not easily resolved. On the one hand, civilian control was seen in Congress as likely to be less efficient and less capable of managing atomic weapons and maintaining secrecy than was military control. On the other hand, six months after the end of the war, the country at large was no longer convinced of the virtues of the military. A strong military voice in atomic energy was less appealing than it might have been earlier. Then, overnight, the situation changed drastically when news came from Canada that bomb secrets had reached the Russian embassy through British physicist Alan Nunn May. When he visited the Chicago laboratory in 1944 as a member of a Canadian atomic energy delegation, he had learned about the research there and about the production of fissionable material at Hanford. Perhaps for the first time, the U.S. public was alerted to the intensity of Soviet expansionist aims. These would separate Eastern and Western Europe by what Winston Churchill called the Iron Curtain and create a long-lasting period of deep contention for physical and ideological global dominance between the United States and the USSR that Churchill described as the cold war. Patterson's cautionary comments grew more meaningful, and General Groves's testimony on security before the McMahon committee made a deep impression. There was nothing that could be done about Russian espionage, but public opinion that had been moving away from inclusion of the military in the affairs of the commission turned back toward that direction.

Opinion on the civilian-military question was not divided along narrow partisan lines. Dwight Eisenhower, the army chief of staff and Chester Nimitz, the chief of naval operations, emphasized their "desire to establish civilian control to the last possible degree of national safety" but also believed that the military services should have a strong voice in matters of national security.[1] Senator Vandenberg, after hearing General Groves, denounced military exclusion in favor of military review of the decisions of a civilian commission. He proposed an amendment to the McMahon bill by which the commission would have absolute freedom to make any decision it wished but the army chief of staff would review any action on military questions. A majority of scientists remained committed to military exclusion as did a majority of citizens' groups, but not all spoke against the Vandenberg amendment.

The differences persisted into April 1946. By then, much of the emotional heat had been dissipated in the stalemate that had developed. May-Johnson had fallen behind because it overemphasized the military and weapons aspects of atomic energy, and McMahon's bill was perceived as too liberal in granting power to a civilian commission without assurance that it could conduct business efficiently and provide for national security. This led to further modification of the McMahon bill: the five civilian members of the commission were retained but in staggered five-year terms, again appointed by the president; they were to be supplemented by a general manager also appointed by the president, which would allow the directors of the divisions to be chosen by the commission.

Those revisions were important, but additional substantive changes that took into account concerns of the scientific community and the military were also introduced. Three mandated committees were incorporated into the bill: a general advisory committee of scientists and engineers for technical matters, a military liaison committee for military matters, and a joint House-Senate committee on atomic energy.

The General Advisory Committee (GAC) not only gave the scientists a consultative voice in atomic energy affairs, but gave the commission a powerful arm to inspect, criticize, and evaluate technical progress under the management of the commission. The GAC would neither vote nor attend commission meetings, but its advice would be sought and, in the event of a severe difference with the commission, the GAC could always appeal to the Joint Committee or even to the president.

The Military Liaison Committee (MLC) was intended to give the armed services the highest-level information on atomic energy. It would give them

FIGURE 3.1. *Top*: The first AEC Commission. *Left to right*: William Waymack, Lewis Strauss, David Lilienthal (chairman), Robert Bacher, and Sumner Pike.

Source: Department of Energy.

Bottom: Members of the General Advisory Committee visited Los Alamos. The picture was taken shortly after the landing at the Santa Fe, New Mexico, airport, April 3, 1947. *Left to right*: James B. Conant, Robert Oppenheimer (chairman), General James McCormack, Hartley Rowe, John H. Manley, Isadore I. Rabi, and Roger S. Warner. Manley was the committee's executive secretary. McCormack and Warner were members of the commission's staff.

Source: R. G. Hewlett and Francis Duncan, *Atomic Shield: A History of the U.S. Atomic Energy Commission*, vol. 2, *1947/1952* (Washington, D.C.: U.S. Atomic Energy Commission, 1972), p. 46.

the opportunity to be heard by the commission and to guide the commission on military matters. Creation of the MLC did not specifically enlarge the role of the armed services in military applications of atomic energy—the MLC would not vote or attend commission meetings—but it reduced the exclusive control of the commission.

The provision in the bill that stated the commission alone would own and operate its plants was changed to specify ownership only and allow the commission to continue the system of contractor operation of the plants that had originated with the Manhattan Project. Similarly, wording that gave exclusive control of the weapons stockpile to the commission was changed to state that the president might direct the commission to deliver such quantities of weapons to the armed services as were deemed necessary. Presumably, the president might do so without the pressure of an immediate emergency. Changes in the section on information control were introduced because the Canadian spy case had heightened the fear of espionage. The distinction between basic scientific and related technical information was eliminated, and the emphasis was shifted to restrict transfer of information concerning military applications. Patent provisions were also extended, and penalties for violations to benefit any foreign nation were stiffened.

Finally, a compromise that went to the heart of the civilian-military issue was inserted to change Senator Vandenberg's amendment. Rather than have the president appoint members of the MLC, the senator proposed that the armed service secretaries do so, which would create a direct line of appeal of commission decisions by the armed services.

With these many changes, the bill went to the full Senate as a triumph of compromise. Secretary Patterson, General Groves, and the armed service heads, General Eisenhower and Admiral Nimitz, regarded the responsibility and authority of the military to be reasonable and satisfactory. Citizen groups that were determined to have civilian control of atomic energy were satisfied and even relieved to have a clearly contained and defined role for the military. Scientists saw two important safeguards firmly in place: The presence of the GAC gave reassurance that understanding and appreciation of the scientific aspects of atomic energy would be available to the commission, the MLC, and the Joint Committee, thus new ideas and new actions would be less likely to get lost as the commission became immersed in managing the atomic energy enterprise. Equally important to the scientists was their freedom and the freedom of the commission to perform research that would take them beyond the scientific threshold they had already crossed. By maintaining leadership in basic research, the United States would retain its leadership and security in the atomic age.

The Senate passed the McMahon bill unanimously on June 1, 1946. No such action could be expected in the House. A protectionist flavor was stronger there, and many members preferred the May-Johnson bill, although they realized that it could not be resurrected. Despite this, the House adopted the McMahon bill with few changes on July 20, by a vote of 265 to 79. It then went to the Senate-House conference committee, from which it emerged essentially in the form it had when it left the Senate. On July 26 both houses accepted the McMahon bill with minor modifications, and six days later the Senate Special Committee witnessed the signing of the Atomic Energy Act of 1946 by the president.

There were many who helped bring to pass the Atomic Energy Act, but the contributions of a few were especially important. Foremost were the bill's authors, Senator McMahon and James R. Newman, who kept their focus on the vital issues and refused to be discouraged. Byron S. Miller, a lawyer who had worked during the war in the Office of Price Administration, might also be credited as an author of the bill for the interviews he conducted and the drafts of the bill he helped to write. Senator Vandenberg was a valuable resource in the Congress, and President Truman entered the argument when it was most appropriate.

CREATION OF BASIC RESEARCH LABORATORIES

The next order of business was for the president to appoint the five AEC commissioners. In the autumn of 1946, however, General Groves continued as the head of the Manhattan Project, the fabric of which, to his dismay, was beginning to unravel. The isotope separation plants were operating more efficiently than before, but demand for their products was on hold. The technical challenge had faded, and everyone, managers and workers, was eager to return to peacetime work. The du Pont Company, for instance, notified the Manhattan Project that it intended soon to relinquish its management responsibilities at Hanford, and Groves was unable to dissuade them.

The situation at Los Alamos was even more serious. The state of high excitement and urgency had passed, and scientists and their families were in a hurry to return to the universities, as were students seeking advanced degrees and jobs. Engineers and technicians, finding themselves in demand by the resurgent peacetime industry, also left the laboratory. The uncertainty of the future of atomic energy during the many months of debate over the Atomic Energy Act had even frightened away some who saw new scientific challenges at Los Alamos. The result was a laboratory with a sadly reduced capability.

Nevertheless, Groves's responsibility to the Manhattan Project appeared to be intact for the near term, and he set about repairing the setbacks with energy and decisiveness. He approached the General Electric Company, which had expressed interest in constructing electric power plants driven by reactors, striking a deal by which GE would take over management of Hanford, with its two plutonium-generating, high-power reactors. In return, the Manhattan Project agreed to finance a laboratory in Schenectady, New York, which would be government owned but used by GE for fundamental research and development of reactors. In addition, Groves persuaded A. H. Compton, who had recently become chancellor of Washington University, in St. Louis, Missouri, to remain as head of atomic energy research in the Chicago area and to oversee a program of reactor development at the Argonne Laboratory in that area, under its director, Walter Zinn.

Although criticized for his strict military outlook, Groves had learned the lesson that research pays. He used the opportunity given him by the

FIGURE 3.2. *Opposite page top*: Laboratory directors of the AEC with the general manager, January 18, 1947. *Front row, left to right*: Frank H. Spedding, Ames, Iowa; Carroll Wilson (general manager); and C. Guy Suits, Knolls. *Standing, left to right*: Ernest O. Lawrence, Berkeley; Philip M. Morse, Brookhaven; Eugene P. Wigner, Clinton; and Walter H. Zinn, Argonnne.

Source: R. G. Hewlett and Francis Duncan, *Atomic Shield: A History of the U.S. Atomic Energy Commission*, vol. 2, *1947/1952* (Washington, D.C.: U.S. Atomic Energy Commission, 1972), p. 110.

Opposite page bottom: David Lilienthal with members of the AEC testifying at hearings before the Joint Committtsee on Atomic Energy of the House and Senate in the spring of 1949. At these hearings, Senator Bourke B. Hickenlooper (Iowa) charged the AEC with "incredible mismanagement," but the Joint Committee's eighty-seven-page majority report vindicated the Lilienthal commission.

Source: R. G. Hewlett and Francis Duncan, *Atomic Shield: A History of the U.S. Atomic Energy Commission*, vol. 2, *1947/1952* (Washington, D.C.: U.S. Atomic Energy Commission, 1972), p. 270.

time interval between the Manhattan Project and the AEC to encourage the establishment of research laboratories. In addition to funding an R&D laboratory for the GE Company and a research program for the Argonne Laboratory, he accepted Ernest Lawrence's request for a subsidy of $170,000 to support completion of the 184-inch-diameter particle accelerator on the Berkeley campus of the University of California. It proved to be one of the first uses of government funds for basic physics research in a university.

Groves also recognized that renewed investment in the Los Alamos Laboratory was necessary if it was to survive as the vital force it had been during the war. The laboratory was ideally located for people fond of living near the high desert, but it needed to make the transition to a peacetime community to attract scientists and their families. Groves arranged for the construction of wells, pipelines, and pumping stations to bring water to a central station and eliminate the queues that resulted when existing water lines had frozen. He persuaded Patterson—Los Alamos was still an army base—to authorize three hundred units of permanent housing. And he directed the new laboratory director, Norris Bradbury, to develop a master plan for the laboratory itself, to replace its previous hodgepodge construction. These actions represented a vote of confidence in the future of the laboratory and continuing research in atomic energy.

Although the laboratory's staff was spread thin, given the responsibility for bomb tests in the South Pacific and work on improved detonation of early-type atomic weapons, the depleted Theory Division continued to study the prospect of atomic bombs made of hydrogen instead of uranium and plutonium. In September 1946 the reported promise of this so-called thermonuclear weapon began to influence planning for the future of the laboratory. At the same time, Groves took steps to move many of the routine operations still conducted at Los Alamos elsewhere, recognizing that the military had to take responsibility for straightforward operations still conducted at the laboratory in order to free the laboratory to do research that only its specialized personnel were trained to do. The scientists welcomed relief from these tasks as they channeled greater effort into improved fission bombs and the possibility of hydrogen bombs.

In early 1946 Groves had appointed an advisory committee on research and development to help him prepare a 1947 budget for the Manhattan Project. The committee consisted of seven members who had figured prominently in science during the war, most of whom had since returned to the university. They recommended that the Manhattan Project expand its activities to include a larger number of institutions, with the aim of developing fissionable materials and power. The committee proposed establishing two laboratories, one at Argonne and one somewhere in the northeastern states. These laboratories, each managed by a board of directors chosen primarily from universities, would be channels through which federal funds would flow to support nuclear research. Creation of a national laboratory in the West was also recommended for a later time. The committee recommended setting the fiscal year 1947 budget for research and development at $20 to $40 million. It also endorsed the distribution of radioisotopes for medical research, particularly for diagnostic purposes and cancer treatment, and recommended that nuclear physics research at Berkeley continue to be subsidized with the understanding that Berkeley—possibly as a special type of national laboratory—would assist other U.S. institutions in the design and construction of accelerators.

The budget that was finally submitted went far beyond the recommendations of the advisory committee. It allocated $72.4 million for research, with 68 percent for construction: $20 million for Clinton at Oak Ridge, Tennessee; $10 million for the laboratory GE was to operate at Schenectady; $9.4 million for the proposed northeastern national laboratory; $5 million for Argonne; and $2.5 million for miscellaneous laboratory construction. The remaining $23 million were reserved for operating expenses for those laboratories and for nine other institutions, all universities except one. The Military Appropriation Act of July 1946 contained this budget and was the first substantial appropriation of federal funds for atomic research in peacetime outside the Manhattan Project.

Groves moved rapidly to create the national laboratories proposed by his advisory committee. The University of Chicago accepted a contract to operate Argonne National Laboratory, and the university and the Manhattan Project approved a statement defining its organization and operating policy that had been drafted by twenty-four participating midwestern institutions. The New York State Board of Regents chartered nine private universities in the Northeast as the Associated Universities, Inc., to manage the Brookhaven National Laboratory, a new fundamental science research laboratory in Long Island, New York. Elsewhere, fourteen universities spread in an arc from the District of Columbia to Texas formed the Oak Ridge Institute of Nuclear Studies to use the facilities of the Clinton Laboratories for basic research, the equivalent of a national laboratory in the Southeast.

The Manhattan Project approved the institute, and it received a Tennessee charter of incorporation.

While this plan for a network of federally supported research institutions was undertaken, on October 28, 1946, the president appointed the five AEC commissioners proposed in the McMahon bill, and on December 31 the commission formally took control of the Manhattan Project, subject to the conditions of the Atomic Energy Act and the oversight of the Joint House and Senate Committee on Atomic Energy. The chairman of the commission was David E. Lilienthal, who had served fifteen years on the Wisconsin Public Service Utility Commission and since 1933 had been chairman of the Tennessee Valley Authority, the federal electric power utility that had been an integral part of Roosevelt's New Deal program of rural electrification. His knowledge of the ways of the federal government and of power technology on a national scale, combined with his leadership ability, made him a natural choice to head the commission. Appointed with him were Sumner T. Pike, a former member of the Securities and Exchange Commission; Lewis Strauss, an independently wealthy financier, who as a navy reservist rose to the rank of admiral and adviser to the secretary of the navy; William W. Waymack, editor of the *Des Moines Register and Tribune*, who was serving as a public director of the Federal Reserve Bank of Chicago and had received a Pulitzer Prize for editorial writing; and Robert F. Bacher, one of the key scientists at Los Alamos, a professor of physics at Cornell University and aide to U.S. representatives to the United Nations. President Truman proudly claimed that he had chosen the five members without knowledge of their political affiliations, and in fact all but the chairman—who was an independent—were Republicans. The commission realized that they could sustain public approval for the difficult decisions they would face only if they remained and were perceived as nonpolitical.

The commission began doing business as proprietors of the atomic energy enterprise without offices or staff of their own, although they did receive a million dollars from the Treasury to pay expenses. Consequently, they were initially dependent on the plans for the future inherited from General Groves and the Manhattan Project staff. The commission had the authority to replace contractors and employees in any commission plant or laboratory with its own contractors and people and thereby to take active control in addition to ownership. Instead, it chose to retain the army's system for dealing with the plants and laboratories it owned and to work with the thousands of scientists, engineers, and technicians already in place in

FIGURE 3.3. *Top*: Builders of the Bevatron. Standing in front of the giant particle accelerator at Berkeley are the scientists principally responsible for its design and construction. *Left to right*: Ernest O. Lawrence, William M. Brobeck, Edward J. Lofgren, and Edward M. McMillan.

Source: R. G. Hewlett and Francis Duncan, *Atomic Shield: A History of the U.S. Atomic Energy Commission*, vol. 2, *1947/1952* (Washington, D.C.: U.S. Atomic Energy Commission, 1972), p. 302.

Bottom: Celebrating a milestone in the construction of the Cosmotron, the particle accelerator at the Brookhaven National Laboratory at Long Island, New York, in December 1950. G. Kenneth Green stands in the center of the group. *Left to right around the circle*: Abraham Wise, George B. Collins, Charles H. Keenan, Gerald F. Tape, M. Stanley Livingston, Martin Plotkin, Lyle Smith (mostly hidden), Joseph Logue, and Irving L. Polk.

Source: R. G. Hewlett and Francis Duncan, *Atomic Shield: A History of the U.S. Atomic Energy Commission*, vol. 2, *1947/1952* (Washington, D.C.: U.S. Atomic Energy Commission, 1972), facing p. 302.

private industry and universities. As a result, the administration of the AEC remained decentralized and flexible.

The commission did not immediately agree with General Groves on decisions involving long-range commitments, particularly the new basic research laboratories, for which there was no precedent. It inspected the organization and location of the proposed General Electric laboratory and the national laboratories at Argonne and Brookhaven, for which funds had already been appropriated, before finally confirming the earlier decisions. This illustrated the boldness with which General Groves could act within the relatively loose constraints of the War Department compared with the initial caution of the commission. The commission found itself on firmer ground after the president appointed a general advisory committee (GAC) consisting of nine experienced leaders in the wartime scientific effort, three of whom had won or would win Nobel Prizes. The initial recommendation of the GAC, with Oppenheimer as chairman, underscored the need to compete in a dangerous international environment. Furthermore, the GAC recommended extending the policy of building new basic research laboratories that General Groves and his advisory committee had advocated, arguing convincingly that investments should be made not only in the bricks and mortar of new laboratories but also in new equipment—especially higher-energy particle accelerators—that would permit physicists to expand the boundaries of their science.

Thus, in the history of U.S. atomic energy, the AEC became the principal funding agency of university and industrial science and technology. It did so not only in nuclear physics and chemistry but also in areas of science peripheral to and in many instances far from those subjects. Because ideas are not enclosed by administrative boundaries, the AEC found itself supporting research in biology, nuclear medicine, and materials science, in addition to the core research in high energy physics.

THE HYDROGEN BOMB

Three years after passage of the Atomic Energy Act of 1946, the United States procured evidence that the USSR had detonated an atomic bomb. This was not completely unexpected because U.S. scientists expected that Soviet scientists would need about the same length of time to achieve that goal as they had. Nevertheless, it came as a shock to realize that the U.S. monopoly of atomic weapons was at an end. Russian possession of the bomb was viewed as an immediate threat to U.S. security. Few in the government believed that the USSR, dominated as it was by Stalin, would refrain from an atomic attack on the United States.

The AEC responded by compiling a detailed inventory of the U.S. atomic arsenal. Was it sufficient in number and power to ensure devastation of the USSR in the event of a first strike against the United States? A partial answer to this question could be made by listing the quantity of fissionable material produced each week and the number of atomic weapons ready for delivery. Answers, however, were also needed to questions concerning more powerful weapons than the WWII type and also small atomic weapons for tactical use. The joint committee, chaired by Senator McMahon, and the MLC pressed for answers to those questions. In particular, they urged progress on the much more powerful hydrogen bomb then under study by the rechristened Theoretical Division at Los Alamos.

That study indicated that a bomb made of hydrogen would have as much as one thousand times the explosive power of the atomic bombs of WWII. The principle on which the idea of the new bomb was based—the fusion or joining of hydrogen nuclei, the same process that generates the energy in stars—was recognized as a thermonuclear reaction, that is, a nuclear reaction brought about by extreme heating of nuclei (as in the core of a star) that would give rise to the emission of a large quantity of energy.

Before the Soviets exploded a uranium weapon, there was no compelling motivation to explore the possibility of a thermonuclear weapon and there were practical reasons to refrain from doing so. The issues involved in proceeding toward a hydrogen bomb were essentially the same as those that had been faced in deciding to make a uranium bomb in 1940. Research on hydrogen bombs would be enormously expensive and impede work on improving fission bombs, just as it was anticipated in 1940 that research on a uranium bomb would use scarce resources and slow the improvement of conventional weapons.

The Soviet explosion of a fission bomb changed everything. Once again, Americans were concerned about being left behind. Scientists at Los Alamos began to explore more intensively the technical questions involved in constructing a fusion weapon. As they reported on their progress, the idea of a bomb of that extraordinary power captured the imagination of the members of the joint committee and the MLC. They saw it as a way to reassert U.S. ascendancy in atomic weapons. Within the AEC and the GAC,

Table 3.2 Members of the U.S. Atomic Energy Commission and the General Advisory Committee

---

U.S. ATOMIC ENERGY COMMISSION

David E. Lilienthal, chairman	November 1, 1946–February 15, 1950
Robert F. Bacher	November 1, 1946–May 10, 1949
Sumner T. Pike	October 31, 1946–December 15, 1951
William W. Waymack	November 5, 1946–December 21, 1948
Lewis L. Strauss	November 12, 1946–April 15, 1950
Henry D. Smyth	May 30, 1949–September 30, 1954
Gordon E. Dean	May 24, 1949–June 30, 1953
chairman	July 11, 1950–June 30, 1953
Thomas E. Murray	May 9, 1950–June 30, 1957
T. Keith Glennan	October 2, 1950–November 1, 1952
Eugene M. Zuckert	February 25, 1952–June 30, 1954

GENERAL ADVISORY COMMITTEE

James B. Conant	December 12, 1946–August 1, 1952
Lee A. Dubridge	December 12, 1946–August 1, 1952
Enrico Fermi	December 12, 1946–August 1, 1950
J. Robert Oppenheimer, chairman	December 12, 1946–August 8, 1952
Isidor I. Rabi	December 12, 1946–August 1, 1956
chairman	October 1952–July 1956
Hartley Rowe	December 12, 1946–August 1, 1950
Glenn T. Seaborg	December 12, 1946–August 1, 1950
Cyril S. Smith	December 12, 1946–January 10, 1952
Hood Worthington	December 12, 1946–August 1, 1948
Oliver E. Buckley	August 2, 1948–August 1, 1954
Willard F. Libby	August 7, 1950–September 30, 1954
Eger V. Murphree	August 7, 1950–August 1, 1956
Walter G. Whitman	August 7, 1950–August 1, 1956
John von Neumann	February 27, 1952–August 1, 1954
James B. Fisk	September 22, 1952–August 1, 1958
John C. Warner	September 22, 1952–August 1, 1964
Eugene P. Wigner	September 22, 1952–November 19, 1956

Source: R. G. Hewlett and Francis Duncan, Atomic Shield: A History of the U.S. Atomic Energy Commission, vol. 2, *1947/1952* (Berkeley: University of California Press, 1972), p. 664.

however, there was greater skepticism because of the technical uncertainties and the large expense. Moreover, they anticipated that the USSR would also move in the same direction and the result would be an arms race in fusion as well as in fission bombs. Would this not, they asked, bring about less rather than more national security? And would not a reasonable alternative to development of thermonuclear bombs be to announce that the United States would refrain from doing so if the USSR would agree to the same? This alternative—occasionally referred to as "to announce to renounce"—was seen by some members of the commission and the GAC, who were appalled by the power for mass destruction, as a realistic step toward a world disarmament treaty.

It was inevitable that the two very different courses of action—development or renunciation—would give rise to deep differences of opinion among members of the commission and the GAC and would alienate them from the joint committee and the MLC, where the consensus was for full-scale development. In the highly charged atmosphere of the time, some individuals on the commission and the GAC saw the pursuit of thermonuclear weapons as sinful and potentially in the same category as the mass slaughter of humanity by the Nazis. Others saw no essential difference between those weapons and the more powerful fission weapons that were already being pursued.

The joint committee had no reservations about the development of thermonuclear weapons and no patience whatsoever with the idea that an understanding with the USSR might be possible. An underlying source of tension between the commission and the GAC, on the one hand, and the Joint Committee, on the other, was the position taken by the joint committee that they and the president were elected to pass judgment on such questions based on the commission's and the GAC's technical advice, not on their ideas of global strategy or canons of morality.

This role was extremely hard for some members of the commission and the GAC to accept. Lilienthal, for one, felt that the United States was making a tragic error in giving up what he and others saw as a unique opportunity to achieve world disarmament. He was convinced that disarmament was in the best interest of the country, which was unscarred by a modern war, and that renunciation was worth a try in spite of the high probability that it would fail. Even before President Truman decided to go ahead with development of the hydrogen bomb, Lilienthal, weary in body and soul from the constant tumult of the previous three years, decided to resign from the commission. At the same time, Lewis Strauss, a strong proponent of development, felt his aim was accomplished when the president assented to this project and so resigned.

The development of the hydrogen bomb would require the construction of new plants for reactors to generate hydrogen isotopes and a new laboratory akin to Los Alamos. The method of achieving detonation of a hydrogen bomb required fresh ideas and difficult experiments to confirm them. There was much disagreement about the experiments and their validity, and the situation was worsened when Communist forces attacked South Korea on June 25, 1950, and the United States entered the Korean War. Some distinguished physicists—among them Oppenheimer—saw the technical problems presented by a fusion bomb as likely to be insuperable. Indeed, without the original, very clever work of Stanley Ulam and Edward Teller at Los Alamos, the hydrogen bomb as a true fusion weapon would not have emerged at that time. In Washington, there was fear of an expanded conflict, possibly another world war, and the urgency of increasing the stockpile of fission bombs and developing thermonuclear bombs intensified. Priorities shifted once again, and the GAC was pushed to estimate the prospect for the successful construction of a hydrogen bomb. The joint committee under Senator McMahon would tolerate nothing less than an all-out effort to produce the weapon. Expenditures that might have been questioned at another time were approved without a second thought. The commission concurred, and progress toward the creation of a hydrogen bomb was rapid. On October 31, 1952, a single hydrogen bomb measured at the equivalent of 10.4 million tons of TNT—or five hundred times the strength of the Hiroshima bomb—was detonated by the United States on Bikini Atoll in the South Pacific.

THE OPPENHEIMER INVESTIGATION

The passions that attended the development of the hydrogen bomb came to a bitter climax two years after the test. The first briefing of president-elect Dwight D. Eisenhower by the AEC was in early November 1952 in Augusta, Georgia. It was a secret briefing, ordered by President Truman, to acquaint the new president with the facts of the hydrogen bomb test a few days earlier. Eisenhower was stunned by the news and immediately inquired about plans to keep strict secrecy. His dealings with the Russians in the seven years since the end of WWII had convinced him that it was best to keep them off balance by maintaining secrecy. If there was an advantage to announcing the successful test of a U.S. hydrogen bomb, then and only then would Eisenhower go public. With that frame of mind, it was natural for Eisenhower to ask Lewis Strauss, who agreed, to serve as his special assistant on atomic energy. Eisenhower's outlook was also shared by most members of the joint committee and particularly by its chairman, McMahon, before his untimely death from cancer at age forty-nine, just prior to the hydrogen bomb test.

There was, however, considerable ambivalence contained in Eisenhower's position. He desired secrecy as far as the Russians were concerned, but he was also aware of the need to inform the American people of what had been done and of the international crises that the nation faced as a result. A panel to study possible U.S. proposals for disarmament had been organized by Dean Acheson, secretary of state under Truman, with Oppenheimer as chairman. The panel produced a report titled "Operation Candor" that circulated within the government, stressing the terrible consequences of an atomic war—a hydrogen bomb war—and outlining the advantages to the United States of seeking disarmament. Eisenhower very much supported "Operation Candor" and cast about for practical proposals that might be presented to the nation and the world based on its ideas. The president in essence sent a mixed message to those concerned with atomic weapons and national security: on the one hand, that it was prudent to build up the nuclear arsenal and, on the other, idealistic to promote disarmament.

Although completely loyal to Eisenhower, Strauss strongly opposed the ideas of "Operation Candor" and indeed to any release of information that might, in his view, benefit the Russians even in the most minimal of ways. Strauss and others who agreed, among them Commissioner Thomas E. Murray, still balked at Oppenheimer's unrelenting opposition to the hydrogen bomb and decided that an attempt should be made to block Oppenheimer's employment in any future consultative capacity for the government. They were backed by a report on Oppenheimer's activities compiled earlier by William L. Borden, then the executive director of the joint committee, who claimed that Oppenheimer was at the very least a security risk. When the USSR tested a bomb using hydrogen as part of its explosive material just nine months after the U.S. test, those suspicious of Oppenheimer asked whether he had used his influence purposely to delay the U.S. effort and allow the Russians to catch up. Still others questioned the advice that had come from other committees he had chaired. For example, Oppenheimer's advice to the Strategic Air Command initially cast doubt on whether hydrogen bombs could be built, and he suggested that its strategy be focused on other weapons.

The situation came to a head when Gordon Dean, on Oppenheimer's request and just a few days before retiring as AEC chairman, renewed Oppenheimer's consultantship to the AEC for the year June 30, 1953, to June 30, 1954. Assuming no opposition, Dean took the step without consulting his fellow commissioners or his successor, Lewis Strauss. This action set the stage for a profound tragedy.

A second report from Borden, by then a private citizen whose own prior record at the joint committee of keeping classified documents secure was far from spotless, went from J. Edgar Hoover, head of the FBI, and Strauss to Eisenhower and members of the National Security Council. It was accompanied by an FBI report on Oppenheimer dating back to 1942. This report was interpreted by some as indicating Oppenheimer's questionable behavior and by others his active disloyalty. Some suggested that Oppenheimer, who was abroad at the time, might be ready to defect. The issue of Oppenheimer's status and his access to classified information was brought to the president as an urgent matter, but Eisenhower took the only action available to him: he delayed it. Attempting to pacify all parties and treat Oppenheimer fairly, he nevertheless suspended Oppenheimer's clearance pending an investigation by the AEC, since the clearance involved concerned Oppenheimer's consultantship.

Soon afterward, Chairman Strauss and the new general manager of the AEC, General Kenneth D. Nichols, who had been General Groves's assistant throughout the Manhattan Project, met with an astonished Oppenheimer to offer him the opportunity to resign in order to forgo an investigation. That was an option that Oppenheimer refused. He feared such action would be a tacit admission that he was guilty of some crime that, as far as he was concerned, he did not commit.

The commissioners arranged for a board of three members from outside the government to review Oppenheimer's situation. The board was headed by Gordon Gray, a lawyer and former publisher, who had been assistant secretary of the army in 1947 and subsequently an assistant to the president of the United States until he became president of the University of North Carolina in 1950. The investigation began on April 12, 1954, and went on, for five full days each week, until May 6 of the same year. The board concluded that

Oppenheimer was devoted to his country but voted two against one that, because of bad judgment, he was a security risk and his access to classified material should be withdrawn. Four of the five commissioners concurred in this recommendation; the only exception was Henry Smyth.

The public—to whom Oppenheimer was a hero—was generally critical of the result of the investigation. The scientific and academic communities were divided: most thought he had been railroaded in an effort to put scientists in their place, but some believed there was justification for the verdict against him. This was the time of McCarthyism and of the Soviet spies Allan Nunn May, Klaus Fuchs, and the Rosenbergs. Emotions ran high, and rational argument ran low. Few if any recognized how similar the fates of Douglas MacArthur and Robert Oppenheimer were, both men of remarkable gifts and accomplishments brought low by their own arrogance. Gordon Gray was disturbed by the authority with which Oppenheimer placed his own judgment, as Gray put it, "over that of more responsible persons."[2] President Truman could have said precisely the same thing of MacArthur. At the conclusion of the investigation, Oppenheimer returned to his position as director of the Institute of Advanced Study in Princeton, New Jersey, where he remained until his death in 1967.

Rescinding Oppenheimer's clearance to receive classified information and serve as a government consultant had many repercussions within the community of American physicists. The investigation had pitted Ernest Lawrence and Edward Teller against Oppenheimer. Younger physicists found themselves passively taking sides, unable to resolve the conflict between their personal loyalties. For example, Robert Serber, who had a position at Berkeley, moved from California to Columbia University, in New York, leaving behind the academic institutions that were home to Oppenheimer, Lawrence, and himself. More generally, scientists were disabused of the notion fostered by WWII that they were indispensable and would always be handled with kid gloves by the government.

In the eight years between the creation of the AEC and the Oppenheimer investigation, atomic energy grew to be perhaps the most important part of the national defense. In that period, scientists had been introduced to the inner councils of government in peacetime and in turn the government had learned a bit about science and a lot about scientists. What emerged was a kind of love-hate relationship. Both sides recognized their mutual attraction and interdependence. It was clear that they would not be able to go completely separate ways.

The navy acted to establish close relationships with university scientists after WWII through creation of the Office of Naval Research.

Not all government science agencies had origins as tumultuous as the AEC's. The Office of Naval Research (ONR) began quietly in the navy and went equally quietly through Congress while the Atomic Energy Act was in the throes of debate.

At the start of WWII, The director of the Naval Research Laboratory was Vice-Admiral Harold G. Bowen, who had previously been the chief engineer of the U.S. Navy. The Naval Research Laboratory was concerned primarily with practical applications that might solve maritime problems. It was the product of a recommendation made originally at the end of World War I by a civilian advisory committee headed by Thomas Edison, and as part of that recommendation the Laboratory reported directly to the secretary of the navy. Admiral Bowen was one of the few high-ranking naval officers aware of the U.S. intent to develop the atomic bomb in 1940 and of the promise of atomic energy for ship propulsion. However, Bowen was forced out of his post in naval research in 1944 because he was difficult and had antagonized the civilian heads of the OSRD by insisting that naval weapons development should be kept out of civilian hands. His inability to work with others was one of the possible reasons that the navy was excluded from the Manhattan Project in the first place.

At the end of WWII, Bowen was determined to bring nuclear power to the ships of the fleet. He had pioneered the use of high-pressure steam power many years before and believed nuclear power would make the fleet largely independent of the land. Bowen looked for a base within the navy from which to pursue his goal and, having learned a lesson from his past experience, concluded that a new agency established to interact closely with civilian scientists might serve his purpose. As director of that agency, his thinking went, he would be in a position within the navy's hierarchy to undertake the development of nuclear propulsion for the fleet.

The road to the agency Admiral Bowen had in mind was not direct. He had been relegated to lesser posts during the war because of his abrasiveness, but his boldness in encouraging innovation was admired by James Forrestal, then secretary of the navy, and by Commodore Lewis L. Strauss, Forrestal's special assistant for research planning and nuclear affairs. Soon after the death of President Roosevelt, Forrestal ordered the creation of the

Office of Research and Inventions (ORI) and transferred to it the Office of Patents and Inventions, the Office of the Coordinator of Research and Development, and jurisdiction over the Naval Research Laboratory and postwar research planning. Admiral Bowen, plucked from relative obscurity, was appointed its chief. He quickly set about cultivating favor with scientists who were returning to universities from the OSRD and the Manhattan Project by offering ORI funds to support their research. The ORI promised freedom of choice and action to the scientists and their universities and was careful to live up to those promises. There would be no burdensome bureaucratic requirements as the price for its support. This was an opportunity too good to miss, and university scientists responded immediately.

Free to concentrate on nuclear propulsion, Bowen had two obstacles to overcome: he needed to convince General Groves that it was in the interest of the army and the nation to grant the navy access to the Manhattan Project, and he needed authorization within the navy for ORI jurisdiction over nuclear propulsion. Direct cooperation with the navy was not welcomed by Groves, but this simply gave added reason for Bowen and his supporters— Lewis Strauss; W. John Kenney, then undersecretary of the navy; and Admiral Luis de Flores, deputy chief of the ORI—to seek congressional authorization for the long-term stability and independence of ORI. President Truman signed the bill creating the Office of Naval Research, the new name for the ORI, without fanfare on August 3, 1946, just two days after signing the Atomic Energy Act.

The second obstacle to Bowen's organizational base for a nuclear navy— authorization within the navy—was too formidable for him to overcome. That issue was settled in favor of the Bureau of Ships, which would later designate Captain Hyman G. Rickover to direct the collaborative effort with the AEC for the development of naval nuclear propulsion systems. Admiral Bowen had failed again to attain a position from which he might direct the conversion of a steam-powered to a nuclear-powered navy. He had, however, been the catalyst that stimulated action by the navy and ironically the choice of Rickover to do that job was a good one. Bowen's legacy to the navy and to a legion of university scientists was the Office of Naval Research, an agency free to act under its congressional charter and the benign neglect of a navy preoccupied with national defense matters. Admiral Bowen went on terminal leave soon after the ONR was created and officially retired a year later.

The ONR staff that Bowen left behind, a mix of naval officers and civilians, was charged to support university-based science in a way acceptable to both academic scientists and their university administrations. Some believed in the need of the nation to have a basic research agency in place while Congress was debating the National Science Foundation; others thought it in the navy's interest to have as allies the scientists whose technical advances had been so important in the war. Still others saw the ONR as a science base for a future organization, possibly a future OSRD. These reasons, coupled with the support of the secretary of the navy, made the ONR an attractive agency to young would-be science administrators. Its first chief scientist was Alan T. Waterman, previously an instructor at Yale University and a member of the OSRD and later the first director of the NSF. Another of its chief scientists, Thomas Killian, became the first chief scientist of the Office of Army Research, and still a third, Emmanuel Piore, went on to become vice president in charge of research at the International Business Machines Corporation.

The ONR satisfied an important criterion specified by Bush: that it be far removed from the operational activities of the navy and independent of the chief of naval operations. It focused on maintaining close relations with universities, university scientists, and engineers by helping them financially to do the science they proposed. Little pressure was exerted to suggest projects or programs of primary interest to the navy. Less than 10 percent of the $86 million received by the ONR in the period 1946 through 1950 was spent on naval science applications. The remainder of the funds went for equipment and the support of basic research in a wide variety of fields. The organizational chart of the ONR in October 1946, shown in figure 3.5, illustrated the level of ambition of its administrators and the method used to create an office staffed at the top by a civilian scientist and a senior naval officer. The laboratories and other divisions on the line just below the chief of naval research were mostly inherited. One revealing item in the organizational chart was the nationwide extent of the branch offices, including a London office that successfully reported on European science and helped promote contacts among American and European scientists.

The ONR used its authority to place contracts with universities by following the precedent established by the OSRD. It paid universities the full costs of research contracts, including indirect costs that went to the university. A scientist who had prepared the research budget was not penalized by deductions for the indirect costs or by overly tight restrictions on expenditures. Again, following the OSRD pattern, the ONR did not advertise for bids

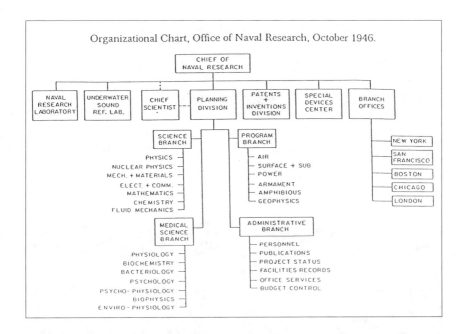

Organizational Chart, Office of Naval Research, October 1946.

CHIEF OF
NAVAL RESEARCH

NAVAL RESEARCH LABORATORY | UNDERWATER SOUND REF. LAB. | CHIEF SCIENTIST | PLANNING DIVISION | PATENTS + INVENTIONS DIVISION | SPECIAL DEVICES CENTER | BRANCH OFFICES

SCIENCE BRANCH
- PHYSICS
- NUCLEAR PHYSICS
- MECH. + MATERIALS
- ELECT. + COMM.
- MATHEMATICS
- CHEMISTRY
- FLUID MECHANICS

PROGRAM BRANCH
- AIR
- SURFACE + SUB
- POWER
- ARMAMENT
- AMPHIBIOUS
- GEOPHYSICS

BRANCH OFFICES
- NEW YORK
- SAN FRANCISCO
- BOSTON
- CHICAGO
- LONDON

MEDICAL SCIENCE BRANCH
- PHYSIOLOGY
- BIOCHEMISTRY
- BACTERIOLOGY
- PSYCHOLOGY
- PSYCHO-PHYSIOLOGY
- BIOPHYSICS
- ENVIRO-PHYSIOLOGY

ADMINISTRATIVE BRANCH
- PERSONNEL
- PUBLICATIONS
- PROJECT STATUS
- FACILITIES RECORDS
- OFFICE SERVICES
- BUDGET CONTROL

FIGURE 3.4. *Top*: Organizational chart of the Office of Naval Research, October 1946.

Source: Harvey M. Sapolsky, *Science and the Navy: The History of the Office of Naval Research* (Princeton: Princeton University Press, 1990), p. 49.

Bottom: Alan T. Waterman, chief scientist of the ONR and director of the National Science Foundation from 1950 to 1962.

Source: National Science Foundation.

but responded to research proposals from scientists. Proposals were ranked by the ONR program officers in the science and medical science branches after consultation with a peer group of scientists in those fields, with support going to those proposals that ranked high. The ONR introduced the idea of providing support for graduate students to help build a future research and academic base. Finally, a minimum of reporting was required by scientists. Since most of the research was unclassified, it could be published in the open literature; a reprint of a published article was often acceptable as a progress report.

FIGURE 3.5. *Left*: Cassius J. Van Slyke, first chief of the Research Grants Office (later Division) of the National Institutes of Health, 1946–1948; director, National Heart Institute, 1948–1952; associate director, NIH, 1952–1958.

Source: Richard Mandel, *A Half Century of Peer Review (1946–1996)* (Alexandria, Va.: Division of Research Grants, National Institutes of Health, Logistic Applications, 1996), p. 21.

Right: Ernest M. Allen, scientist director, U.S. Public Health Service; assistant chief, Research Grants Office, 1946–1950; chief, Division of Research Grants, 1951–1960; NIH associate director, 1960–1963; grants policy adviser, Office of the Surgeon General, 1963–1968; deputy assistant secretary for grants administration, HEW, 1969–1973.

Source: Richard Mandel, *A Half Century of Peer Review (1946–1996)* (Alexandria, Va.: Division of Research Grants, National Institutes of Health, Logistic Applications, 1996), p. 57.

The ONR took responsibility for a number of contracts vacated by the OSRD, servicing seven hundred contracts by 1950. It made itself especially useful in funding expensive, large-scale equipment for universities and took the initiative in funding elementary particle accelerators built on university campuses to explore the burgeoning new field of high energy physics. Soon, however, the ONR found itself overcommitted financially and turned for help to the fledgling AEC. A joint ONR/AEC program began, and it proved to be an extraordinary example of cooperation between two newly formed agencies. Within a few years, high-energy accelerators known as synchrocyclotrons were constructed at the Carnegie Institute of Technology (now Carnegie-Mellon University), the University of Chicago, and Columbia, Harvard, and Rochester universities, each costing several million dollars. The original synchrocyclotron was designed and built at the University of California in Berkeley with the assistance of funds from the Manhattan Project. These accelerator laboratories established U.S. physicists at the forefront of that field and served thereafter as the training grounds of several generations of physics students.

The ONR played a vital part during the birth of government support of basic science and scientists after WWII. This was a period in which discoveries of new phenomena gave rise to new fields of science that needed encouragement, freedom of action, and the financial support that the ONR provided so well and sustained through a critical time. The ONR acted selflessly, operating apart from the mainstream of naval operations even during the Korean War, and was willing to seek help from the AEC to carry out its mission. It was respected and prized by scientists for the effective way it conducted its business and later venerated for the farsighted precedents it set for the future of the science establishment.

After many years of determined independence, the medical community overcame its fear of peacetime affiliation with the federal government.

Before the twentieth century, the U.S. medical community had a tenuous, arm's-length relationship with the federal government. This suited physicians, who were averse to any government interference that might come between them and their patients, and allowed the government to avoid entering a province reserved to the individual states. Nevertheless, there were areas in which the government, at first reluctantly and then more willingly, served a useful function in the field of medicine. It occasionally established a temporary agency to amass data and make recommendations with respect to sanitation, health reform, and control of infectious and chronic diseases. As medical science advanced, there emerged from this tentative beginning an increasing presence of the federal government in epidemiological studies and public health legislation. Thus the origin and evolution of the modern medical science agency, the National Institutes of Health, was the result of a gradual, trial-and-error activity over many decades that caused hardly a ripple of excitement in Congress or in the medical community.

Congress's ambivalent attitude toward medical science is shown by two early inconsistent actions at the end of the nineteenth century. The Marine Hospital Service (MHS), later renamed the Public Health Service (PHS), sought but did not receive a supplementary congressional appropriation to investigate the 1878 yellow fever epidemic in Mississippi. A year later, Congress began a far-sighted but short-lived experiment involving a centralized medical agency, the National Health Board, whose function was to provide funds for yellow fever research to university scientists. The situation was stabilized to some extent in the early twentieth century, after MHS bacteriologists successfully fought epidemics of plague in San Francisco and Hawaii and yellow fever in Cuba. The MHS Hygienic Laboratory was directed to serve as the national clearinghouse for medical scientific information, and the MHS surgeon general was authorized to coordinate the states' public health efforts. Despite these early federal actions, the funding of medical education and research prior to WWI was dominated by private foundations: the Rockefeller Institute and the Carnegie Institution of Washington. These two largely funded research within their own institutions, often called intramural research.

The ravages of the battlefields of WWI strengthened the government's resolve to aid medical research. Congress chartered the National Research Council (NRC) in 1916 to award contracts for "Military-Medico" projects to academic researchers. The NRC Medical Division consisted of fifteen representatives of scientific societies, six or eight members at large, and special committees of outside experts to help make awards, adjust budgets, and administer projects. This movement toward peer review was a departure from the traditional practice of providing funding to only the most experienced investigators, based primarily on their reputations. In 1918 the NRC was extended into peacetime and established as a permanent advisory committee to the federal government. With its annual budget of $800,000, the NRC dispensed 140 fellowships (not all in medicine) and after 1930 began a small grant program.

In the struggle to provide federal relief and recovery programs in the New Deal era of the 1930s, PHS Surgeon General Thomas Parran managed to get $2.5 million for public health research into the 1935 Social Security Act. Soon after, the PHS's Division of Scientific Research began awarding grants to academic scientists, known as extramural grants. The authority to make grants had been placed with the National Advisory Health Council (NAHC) at its creation in 1930, when a congressional act converted the Hygienic Laboratory of the former MHS into the National Institute of Health (NIH). The issue of how grant proposals to the NAHC were to be made and how they were to be evaluated rose again, this time with the assistant surgeon general, Lewis R. Thompson. Thompson was chief of the PHS's Division of Scientific Research and spoke in favor of medical subcommittees composed entirely of individuals outside the PHS who would recommend action based on scientific worth. By 1937 the elements were in place to promote concentrated attacks on single major diseases, as well as general support of research in nonfederal institutions, including fellowships for advanced study. Together, the mortality rate of troops in WWI and the postwar infant mortality prodded Congress to these actions. In the spring of 1937, the NAHC approved $31,520 for new projects. Later that year, Congress created the National Cancer Institute (NCI) and empowered the NCI council to make research grants-in-aid to cancer studies and to make loans of radium to hospitals and university medical centers. In the period from 1938 to 1940, the NCI received 137 applications for funds and awarded $200 million in extramural grants. Almost simultaneously, the OSRD was formed with the Committee on Medical Research (CMR) as one of its divisions.

At the time of the Allied invasion of Europe, three interconnected agencies were contending for control of federal funding for medical research after the war: the surgeon general of the PHS; the CMR, whose leadership included the NIH director; and the NRC's Division of Medical Research. The interlocking memberships of these committees and institutions kept the battle for control from getting out of hand. At the same time, the influence of Vannevar Bush, who was already looking toward the creation of a national research foundation to replace the CMR in peacetime, promoted a medical community free of nonmedical control. So strong was the desire for independence that one unrealistic suggestion expressed by a CMR advisory panel requested control of federally funded medical research without fund management by any federal agency.

Despite the controversy, by the summer of 1946 the residual university contracts in biology and medicine of the CMR were effectively transferred to the PHS. With its adjunct NIH, it took the position held by most of the medical community that research was inseparable from treatment and education. Equally important, the OSRD/CMR/NRC experience from WWII was also transferred. Federal support for biomedicine and medical research was firmly established as the vital prerequisite for the future of national health research and education systems. Furthermore, peer review was recognized as the effective means of assuring quality in research and satisfying the different interests of individual scientists, universities, federal agencies, and federal policy makers. Herein lay the origin of the extramural award system of the NIH, particularly its separation of review and management functions. Rules concerning the behavior and performance of outside consultants— no compensation except for documented expenses and no participation that might involve a conflict of interest—were products of wartime experience.

Of course, consensus did not mean complete agreement. Basic problems remained to be solved. One was inherent in the extensive nature of medicine, which requires supervision by a large number of divisions, committees, and subcommittees. Human afflictions and the basic biomedical sciences needing study have long been legion, and the number of researchers eager to pursue those studies has been nearly as large. This raised difficult questions of principle and organization. How could the vast number of diverse research proposals—good, bad, and indifferent—be handled fairly, competently, and promptly while adhering to the guiding principles? And how would narrow fields of research be satisfied? Moreover, how would the proper balance between the intramural and extramural programs of the PHS/NIH be devised? All agreed, however, that, first, the biomedical knowledge then available was inadequate to mount a direct assault on major diseases; second, that NIH support of broad, far-ranging research programs in the fundamental biomedical sciences was required; and, third, that the way to ensure high-quality sponsored research was to demand that it meet rigorous standards determined by expert peer review.

With the fate of the National Research Foundation yet in the hands of Congress, the director of the NIH, Rolla Dyer, set up a small section primarily to administer the forty-two outstanding OSRD contracts accepted as NIH grants and to carry out those duties as a sponsor of extramural research. The section was headed by Cassius J. Van Slyke, a senior surgeon from the Venereal Disease Division of the PHS, who drafted the regulations under which NIH would exercise "only the most minimal supervision"[3] over extramural researchers. Technical review of proposals would be delegated to advisory panels drawn from universities and medical schools. The NIH Research Grants Office (RGO) opened on January 1, 1946, with Van Slyke as chief and Ernest M. Allen, also from the Venereal Disease Section of the PHS, as his deputy. They were located in temporary quarters pending availability of a permanent location, with borrowed army office furniture and one secretary. It was a beginning much like that of the AEC, coincidentally on the same day.

By August 1947 the tiny RGO had been transformed completely as an organization. A year earlier, the National Mental Health Act had been passed. It authorized $10 million in grants to the states for facilities and research projects, the first permanent legislation authorizing grants to educational institutions to train health care manpower. One result of the act was to increase applications for grants and, as a consequence, to consolidate all five PHS divisions with research grant authority into one single office, the RGO. Soon thereafter, the RGO achieved division status within the PHS and became the Division of Research Grants (DRG), with administrative supervision over all programs of research grants-in-aid of the PHS. Of all the institutes, centers, and divisions (ICDs) that constituted the NIH, the DRG is the second oldest and the division most closely tied to the NIH's Office of the Director. The DRG became the focus of all activities of the extramural system of the NIH, receiving grant applications, assigning them to ICDs, organizing reviews by scientists, managing the information that tracked applications and the progress of awards, and reporting to Congress on critical concerns. The issue of research awards to individual researchers rather than to research groups remained to be settled over time and through the experience of the changing study (review) sections. The emphasis on extramural awards, however, established in the early days with formation of the RGO, did not change significantly. The instinct for autonomy ingrained in the academically educated medical community would ensure that the resources for research would be dispersed widely and not concentrated in a single giant federal medical establishment.

In 1947–1948, the DRG received 2,078 applications, of which 1,526 were approved. The FY 1951 budget for the NIH was $15.75 million, which was then

FIGURE 3.6. *Top*: NIH director Rolla E. Dyer (*seated center*), DRG chief Van Slyke (*seated, fifth from right*), and deputy chief Allen (*standing, far left*) with study section representatives, about 1947.

Source: Richard Mandel, *A Half Century of Peer Review (1946–1996)* (Alexandria, Va.: Division of Research Grants, National Institutes of Health, Logistic Applications, 1996), p. 29.

Bottom: Surgeon General Thomas Parran, USPHS, a key player in the formation of the National Institutes of Health.

Source: Richard Mandel, *A Half Century of Peer Review (1946–1996)* (Alexandria, Va.: Division of Research Grants, National Institutes of Health, Logistic Applications, 1996), p. 18.

decreased by 9.1 percent in a period of national retrenchment during the Korean War. It was thus a slow beginning, but by 1955 the budget of the NIH passed the $79 million mark.

The business of PHS/NIH was initially considered too specialized to warrant technical meddling by Congress in those early years. Perhaps more important, the goals, awards, and research itself of the PHS/NIH medical research system were delivered in terms the average citizen could understand. Moreover, the medical community was conservative and would not tolerate spending on the study of fringe ideas. No questionable medical research awards provoked righteous indignation in Congress. The one matter on which it took a definite position—namely, a broad, nationwide distribution of research funds and medical construction—was largely addressed by the widespread distribution of medical schools and hospitals. Where there was a geographical lack or weakness, the NIH stepped in with a remedial institutional grant.

There was, however, an area in which NIH succumbed to congressional pressure: it involved loyalty tests of grantees and revocation of awards without due process. As early as mid-1947 the NIH was required by the Federal Security Agency (FSA) to secure an affidavit of loyalty from "every incumbent employee."[4] The impetus at the time came in part from incidents at the University of Washington, where three tenured professors were fired for past association with the Communist Party, and in the AEC, where Senate investigators allegedly found security risks. The NIH director complied with the directive for regular employees but deferred to the independent PHS division, the DRG, on the question of NIH fellows and awardees. The DRG insisted that fellows and awardees were not employees and therefore not subject to government investigation or oath requirement. It argued that academic freedom should be respected, that universities held jurisdiction over their faculty who were fellows and awardees, and that loyalty oaths and investigations would lead to serious complications in the award system. This position was not firmly upheld by either the NIH or DRG, which gave way under pressure from the Senate. The division then required from all grant applicants an anti-Communist oath as a condition of any award, and the surgeon general soon extended that requirement, in addition to FBI clearance, to the study section consultants if their service exceeded ninety days.

If loyalty checks were mandatory for federal service personnel, then the PHS had to enforce the ruling for all its employees. The issue of loyalty oaths and investigations for NIH grantees was another matter, and it was fought bitterly by chiefs of the DRG, who capitulated only, as they thought, to protect the newborn NIH from complete disintegration. They saw loyalty oaths as a foretaste of the future, and they were proved right only a few years later, when McCarthyism drove rampant allegations of disloyalty against NIH award recipients. By then, however, the NIH was in no position to resist. President Eisenhower, whose instinct was to wait and see how much irrational, harmful behavior the public would tolerate, allowed his secretary of health, education, and welfare (HEW), Oveta Culp Hobby, to institute a policy denying support to grantees about whom the FBI had some form of derogatory information. Approximately thirty researchers were removed from their projects in mid-1954, among them, the famous, gifted Linus Pauling.

By August 1955 the nation tired of witch-hunters, Marion Folsom succeeded Oveta Culp Hobby as HEW secretary, and the NSF now stood firm to protect its awardees, a position endorsed by Eisenhower. The infamous Senate inquisition of the army had destroyed McCarthy and his Senate colleagues. This episode left its debilitating effects and personal anguish everywhere. It is summed up in a statement by C. J. Van Slyke, who, as NIH associate director for extramural affairs and a pioneer organizer of the Division of Research Grants, was responsible for terminating a number of grants between 1952 and 1955. Van Slyke's statement, given in an oral history interview in 1963, is a cry from the heart for those caught up in the ugly episode:

> Everything ran along fine until McCarthy started acting up, and then we would get instructions from our security officer that this grant headed by scientist number x, or a b, or whatever he was, would have to be terminated. Well, that of course would stop the research work and throw the whole team out of support just overnight, because they had to be stopped immediately. . . . I was the S.O.B. who said, "If you will wire me today that you would change investigators—and I couldn't tell why, I was not permitted to tell them that he was the subject of security questions—and you'll have to recommend somebody else. . . . " I swear I did that dozens of times. . . . We lived through those awful days of McCarthy influence without anything, save the protection I was able to give the research grant program from my desk. I can tell you a good many times I felt like chucking the job. I felt so unclean.
>
> A fellow had signed a petition or something, or had contributed two or three dollars to some cause. This just happened to be causes I would

have contributed to, if I had any money. I would have thought that contributing to free Spain to get rid of the dictator, Franco, would have been a good thing because I'm opposed to dictators. . . . It was these kinds of people who got into trouble. . . . It was the most unfair sort of thing, and it wasn't until Mr. Folsom came in as Secretary that it stopped.[5]

The creation of the NIH and its evolution during the ensuing three decades was remarkable. It was a period in which the Korean War began and ended, the cold war was moving to its height, and the medical community and successive administrations and Congress had to agree on one unprecedented action after another that would sooner or later affect the lives of generations of Americans. Actions agreed on within the medical community broke with the past in trading their independence for the well-equipped, better-staffed, more productive laboratory that favorable peer review and federal funds could buy. In pre-WWII American medicine, this trade-off had few proponents. The majority of the community saw it as government control of medicine. Congress reflected the same view from the government side. It wanted no part of the responsibility for funding medical research or subsidizing medical education at the national level. These positions were not changed by discussion; they simply proved themselves to be irrelevant, given the amazing achievements made during WWII.

The development of the PHS/NIH system with its centerpiece, the Division of Research Grants, was typical of the medical profession of the time. They were determined to be fair and scandal-free and to hold to the guiding principle of peer review as the basis for awards, which in the main freed them from political interference. It was only a matter of time until the nation at large recognized the value of the institutes it had spawned. This occurred not long after the organization had taken on a semblance of permanence, when the Salk vaccine for poliomyelitis was developed in 1955. It is difficult to reconstruct today the dread this disease produced. In the summer of each year, parents of small children grew fearful, since this was apparently when the children were most susceptible. The dread was almost palpable, stimulated by the publicity that President Roosevelt's haven for polio victims in Warm Springs, Georgia, received each year through the March of Dimes. Americans breathed a collective sigh of relief when a vaccine was available to defeat this enemy. The accomplishment lost some of its luster when 204 new cases of polio were discovered among the 400,000 children who had been inoculated with NIH-licensed vaccine manufactured by a commercial company. Fortunately, the NIH and the Communicable Disease Center moved quickly to do its own rigorous safety tests of vaccine lots and were able to restore public confidence in the immunization program. After this experience, NIH field testing of a medicine on a large scale was no longer contracted to outside drug companies or private foundations.

By 1955 there was no question that what was initially an unproven expenditure of public funds for biomedical research and education was subsequently justified many times over.

Legislation creating the National Science Foundation was finally passed in 1950.

As early as 1942, the New Deal stalwart, Harvey Kilgore, introduced a bill to the Senate aimed at creating a national science foundation. When it failed to pass, he reintroduced it in 1943 and again in 1945. Senator Kilgore foresaw the need for a science agency that would support through grants and contracts both basic and applied science research. His foundation included the social sciences and required research funds to be dispensed according to a pre-scribed formula to achieve an equitable geographical distribution. Kilgore's agency was directly responsible to the president and the Congress through their authority to appoint and remove members of its management.

Vannevar Bush agreed on the need for a science foundation but dis-agreed with the strong populist flavor present in Kilgore's vision. In partic-ular, Bush recommended supporting only the best basic research in col-leges, universities, and research institutes; that research was to be identified by critical review of proposals prior to funding. Bush was strongly opposed to any formula for a geographical distribution of funds. He wished to keep the foundation as free as possible of political authority, even going so far as to remove appointment of the foundation's director from the president's purview. Bush thought that support for basic science research in the phys-ical and mathematical sciences was the hardest to acquire and maintain at a steady level. Consequently, he excluded both applied science and social sciences from his proposed foundation because he felt that they were able to attract support separately in their own right.

Bush's ideas were presented to the Senate in a bill submitted by Senator Warren Magnuson on the same day that the Bush report was released by the White House. The stage was set for a contest between the Kilgore and Magnuson bills, with the White House favoring the former because it contained the requirement that the foundation be directly responsible to political authority, namely, the president. After two years of debate, much of it over-shadowed by the stirring drama of the creation of the AEC, Congress sent a bill to the president that he promptly vetoed because it did not provide for presidential appointment of the foundation director and advisory board. The language of the vetoed bill finessed two other issues that divided the Kilgore and Magnuson visions of a foundation. The new agency was instructed to avoid undue concentration of its funds to prevent an overly unbalanced geographical distribution, and the term "other sciences" was added to the proposed list of sciences to allow for the inclusion of social and applied sciences at a later time.

The bill that the president finally signed in 1950 gave him executive con-trol and was a reasonable charter for an enterprise that was completely new to the Congress. By that time, the AEC, ONR, and NIH had several years of experience sponsoring research of university scientists, but it was research that evolved more or less naturally from the functions of their parent serv-ice agencies: the weapons function of the AEC, the modernized navy, and the Public Health Service. The National Science Foundation (NSF) was intended to stand alone, with no attachments or obligations other than to support scientists to carry out the research they proposed. Evaluation of their work would be conducted by fellow scientists and no tangible end product was required. Consequently, the National Science Foundation was seen as a departure from the norm by a Congress intent on exercising detailed oversight of the agencies it created. No wonder that Congress took five years before it would approve the venture.

The agency that was created differed from Bush's report in two impor-tant respects, apart from the change of name from National Research Foun-dation to National Science Foundation. Two of the divisions—Medical Research and National Defense—in the original plan were omitted in the enactment of the NSF. There were several reasons for this. The five-year interval between the end of WWII and the creation of the NSF opened a window of opportunity for existing agencies to satisfy what they rightly regarded as important national needs that they were mostly qualified to fill. The AEC and the ONR had moved quickly to support physical and mathe-matical scientists and were therefore well established before the NSF could get started. At the same time, the surgeon general, representing the Public Health Service and much of the medical community, was reluctant to see biomedical research relegated to a single, untried agency. It was likely to be separated from clinical practice, which would be anathema to that community. This reaction had been anticipated in the report of the Committee on Medical Research that was appended to the Bush report. There, the CMR agreed with the need for a medical research agency independent of the other disciplines under the umbrella of the NSF but hinted that an agency separate from the NSF was also required. But the CMR did not explicitly break ranks with Bush on this issue in 1945. The young, single NIH, however, had an established relationship with the Public Health Service and was the designated heir of the CMR and its wartime contracts. It was structured to permit rapid growth by the accretion of related institutes that would focus the efforts of medical researchers and practitioners on both the research and clinical aspects of diseases.

The reasons for deletion of the Division of National Defense from the NSF were less direct but not mysterious. Influenced by the OSRD's successful program of wartime research and development, Bush recommended a division within the NSF that would conduct long-range research on military problems. Research on the improvement of existing weapons could be done best within the military establishment, but research involving application of the newest scientific discoveries for military needs would be done better by civilian scientists in universities and industry. Both kinds of military research could go forward side by side, and a close liaison between the two could be achieved. Bush emphasized the value of a broad independent program of basic research and that a healthy interaction between military and nonmilitary research would benefit civilian military research. Doubtless, he felt the same way about the Division of Medical Research. But the absence of an NSF or its equivalent immediately after the war left the military uneasy. The armed forces moved to fill the gap by direct contacts with university scientists they knew from the war whose peacetime research was now funded by the newly created AEC and ONR. And, again, they did not relish the idea of a civilian science agency independent of military control

Early in March 1951 President Truman nominated Alan T. Waterman to be director of the NSF. Waterman was formerly the chief scientist of the Office of Naval Research and an alumnus of the OSRD. He was instrumental in forming the ONR policy of funding basic research in universities and helped to negotiate the joint ONR/AEC agreement to do so in 1948. He was welcomed by a congressionally authorized budget of $150,000 for the NSF's first year (1951), although he had expected a number closer to the $15 million upper limit established by the founding legislation. The low budget was in part an aftereffect of the expenditures for the Korean War and in part a reaction to the larger budgets of the other already established science agencies.

Nevertheless, Waterman proceeded actively to bring the NSF together. He recruited a staff from the ranks of the ONR and universities, and, following the pattern of the OSRD, he moved the foundation frequently from one address in Washington to another as it grew. He quickly organized three of the four mandated divisions: mathematics, physical, engineering, and "other" sciences; biological science; and scientific personnel and education. The fourth, medical science, was held in abeyance because the NIH was already funding many of the proposals in biomedical sciences. A small nonclinical medical science program in the NSF was eventually absorbed into the division of biology.

The NSF initiated a nonrestrictive project grant system to respond to proposals. Following the pattern of the OSRD and the ONR and the precedents of the Public Health Service and private foundations, it moved to support the best research within as comprehensive a program as it could afford. A proposal was first submitted by an individual scientist to his or her own institution for more or less pro forma approval, after which it was sent to the NSF. The grant, if obtained, was awarded to the institution, not to the individual, to fix fiscal responsibility. Grants covered the direct costs specified in the proposal plus an additional 15 percent for indirect (overhead) costs. Proposals went to a program officer in the appropriate foundation division who was the scientist's direct, personal contact with the agency. Program officers read each proposal and arranged for their external reviews. The basis of selection of a proposal for funding was peer review. Reviewers from outside the NSF were asked to evaluate the proposal for originality, interest, feasibility, and cost.

Questions of taste and differences of interpretation of the criteria made reviewing less than an exact science. Nevertheless, peer review succeeded in choosing far more good and excellent proposals for support than mediocre or poor ones. The process also kept serious disagreements to a minimum since active scientists were judged by other active scientists. On the other hand, it opened the NSF to criticism on two fronts. The first revolved around the issue of elitism, which has always plagued programs based on peer review. The charge was that a chosen few were responsible for selecting another chosen few in a process that resulted in an exclusive, self-perpetuating network. It was argued that the network was difficult to broach by individuals from institutions with smaller reputations or interests different from the mainstream of research or by those who were less sophisticated in writing grant proposals. The second criticism had to do with the concentration of approved proposals among a relatively small number of universities whose faculty members were an integral part of the network. Some critics believed that this negated the intent of Congress to make the NSF a national agency. Purposely or inadvertently, they said, the system put an obstacle in the way of any university from the "wrong" part of the country that was seeking to improve itself in science, rather than providing encouragement to do just that.

The NSF acknowledged that there was some legitimacy to both criticisms. It established institutional grants to universities for the purpose of improving their status in different areas of science and began fellowship and traineeship awards for postdoctoral scientists and graduate students that allowed recipients to choose where they wanted to study. Many did go to the few outstanding universities, but many came from states throughout the union, and most states had at least one institution that attracted award holders. To a significant extent, this satisfied the desire for a wide geographical distribution of NSF funds.

A serious threat to the integrity of the NSF during its early years, as in the case of the NIH, arose when Senator Joseph McCarthy embarked on his nefarious and ill-fated Communist witch hunts. The Science Board of the NSF, unlike the NIH, elected to hold the line that awards of research grants would continue to be based on the competence of scientists and the merit of their proposals. No security checks would be required for prospective grantees, the board stated, because the agency supported only unclassified research and its awards were made to institutions, not directly to individual scientists. This courageous stand, taken in 1954, the same year that Oppenheimer's security clearance was rescinded by the AEC, worked and helped to protect the NSF and its award recipients. Two years later, President Eisenhower extended the NSF's policy throughout the government.

Another issue of the 1950s, primarily intellectual but with political overtones nevertheless, was concern for support of the social sciences—the "other" sciences in the NSF charter—and how to give them a proper place in the foundation. There was strong opposition to a social science division in the NSF but equally strong pressure for inclusion of some social science funding. Both sides had good reasons: those opposed alleged the difficulty of evaluating the quality of social science proposals and the waste that would be incurred from funding poor proposals; those in favor cited the fact that the social sciences were assuming an increasingly important place in American life and needed more and better study. The initial steps in the direction of inclusion were compromises. Anthropology, human ecology, and demography, all partially quantitative disciplines, were placed in the division of biological sciences, and proposals for research projects in those fields were reviewed. By 1955 a program of sociophysical sciences—mathematical social science, human geography, economic engineering, statistical design, and the history, philosophy, and sociology of science—was inserted into the mathematical, physical, and engineering division for the same purpose. By 1958 the board created an office of social science that brought all the social science disciplines together as parts of a concerted single research effort.

In 1956, the year before Sputnik, the NSF's appropriation was $40 million, which represented the growing respect of Congress for this obvious national asset.

As peacetime science research expanded, Congress acted to empower federal departments to conduct research consistent with their missions.

The four science agencies—the AEC, the ONR, the NIH, and the NSF—were required to encourage and financially support research in U.S. universities. This, however, did not fully satisfy the needs of long-established government departments for better understanding of the technical features of their missions. Government officials were aware that research in certain areas might improve the quality of service to the nation and possibly provide valuable new products and technologies. As a result, proposals to carry out research related to broadly defined missions were submitted to Congress, which approved them readily. These led to the establishment of new, permanent federal facilities, laboratories, and research stations whose purposes were to acquire data on long-term trends in phenomena that influence daily life.

A few examples indicate the direction of this activity. Early in 1948 the secretary of agriculture was authorized to establish laboratories for research and study of foot-and-mouth disease and other animal diseases that constituted a threat to the U.S. livestock industry. These laboratories complemented the private and state veterinary schools and were responsible for legislation to protect the public against diseases that beset both the animals and the humans who handled or consumed them.

In mid-1948 the Weather Bureau in the Department of Commerce was directed to study the causes and characteristics of thunderstorms, hurricanes, cyclones, and other atmospheric disturbances. Over the years, this research has led to better understanding of weather phenomena and to greater accuracy in predicting weather patterns.

A year later, funds were set aside for construction of new facilities for the National Advisory Committee for Aeronautics (NACA), including $10 million for wind tunnels at universities. These studies of flight were the basis of design for the jet aircraft that would traverse continents and oceans a decade later.

In July 1952 a congressional act authorized the secretaries of the army, navy, and air force to establish advisory committees and appoint part-time personnel necessary for research and development activities and to make five-year contracts with extension rights to carry out this program. The objective was to facilitate the performance of research and development in the armed services, but the authorization made allowance for participation and funding of university scientists and engineers, who were encouraged to engage in research only loosely related to the broadly defined military missions. The army and air force created new agencies to do this: the army through the Office of Ordnance Research (OOR) and the air force through the Air Force Office of Scientific Research and Development (AFOSRD); in most respects, these followed the precedent set by the ONR.

The agencies that emerged as auxiliaries to long-established government departments attracted young university scientists as trainees and older scientists as permanent staff. Their laboratories, like those of major industries, grew into an integral part of the scientific resource of the nation.

At first sight, the activity of the decade 1945–1955 appears to have been the product of infatuation with scientific research as the solution for the myriad problems facing the nation. It resembles a romantic interlude run wild, but when the individual actions by Congress are studied and the results evaluated after many years of experience, it is hard to find fault with either the early fascination with science or with its implementation. In short, the idea of funding science in universities and encouraging research in government agencies and industry was a good one. And the nation benefited.

CHAPTER 4

Marriage: 1955–1965

Shocked by Sputnik, the United States created the National Aeronautics and Space Administration

The 1950s were marked by the cold war, a period of tenuous nonaggression between the world's superpowers: the USA and the USSR. At its core was the mutual fear that either side could use its growing arsenal of atomic weaponry to destroy much of civilization. Early in that period, in 1950, the tension in Korea grew into a localized "hot war" that pitted the United States and the Republic of Korea against Soviet-armed North Korea and the Peoples Republic of China. The Korean War was a conventional war; no atomic or thermonuclear devices were deployed. However, many American lives were lost, mostly because of the precipitous demobilization of the U.S. forces after WWII, which left the United States largely unprepared to mount even a limited war in Asia.

The war ended in a stalemate in 1952, the same year that Eisenhower succeeded Truman in the White House. Both sides were back where they had started: the United States and Republic of Korea on the southern side of the thirty-eighth parallel and the North Koreans and Chinese on the northern side. The United States was satisfied that it had thwarted the North Korean attempt to overrun and conquer South Korea and avoided an atomic weapon confrontation with the USSR. The Korean War showed how acute the cold war had become. But U.S. military might, combined with simultaneous U.S. development of the hydrogen bomb, could hedge against the continuing threat presented by the USSR.

The science establishment that was created under the Truman administration had not been called on for special contributions during the Korean War and, apart from budget cuts and the manpower draft that were required in all areas of U.S. society, was left mostly untouched. Three years after the war's end, however, scientific research and development were flourishing. Americans realized that they had an affinity for science, for the excitement and drama as well as the practical benefits. A long history of colorful inventors—Samuel Colt, Eli Whitney, Thomas Edison, and Alexander Bell, to name only a few—had conditioned Americans to view science, both applied and basic, as another manifestation of the pioneer spirit. Absorbing novels and nonfiction accounts of scientists and their work, such as those by Sinclair Lewis, Paul de Kruif, and Hans Zinsser, also influenced several generations of young Americans. The achievements of science and technology in WWII were alive in the minds of the public and Congress.

The uneasy respite over Korea came to a jarring halt with the launch of *Sputnik*, the Soviet Union's successful entrance into space in October 1957. The impact of that event can hardly be imagined now. The U.S. public and Congress, as well as the military, had visions of heavily armed satellites, perhaps with nuclear warheads, with the ability to destroy military bases at home and abroad, leaving cities defenseless against an unassailable enemy. It was likely to be a time enormously more dangerous than the brief era of the flying bombs in WWII. But possibly most shocking to everyone—even those less alarmed by the military potential of *Sputnik*—was the notion that something was wrong with the American system, that it had been outthought and outproduced in an absolutely vital matter. It now appeared that the United States had been unaware that it was being outstripped in the technology of space. How, the country asked, could we have fallen so far behind when our strength should have made us dominant?

An immediate reaction was needed to remedy the situation. But this was easier said than done. *Sputnik 1*, the first man-made satellite in orbit, weighed 183 pounds; the American plan had been to start with the navy's *Vanguard* satellite, at 3 pounds, and work up to 22 pounds in later flights.

That alone was bad news, and more was still to come. The USSR launched *Sputnik 2* less than a month after *Sputnik 1*, and *Sputnik 2* carried a dog as a passenger in addition to its own weight of 1,100 pounds. Then, on December 6, 1957, the much-advertised 3-pound *Vanguard* test vehicle collapsed in flames a few feet above its launch platform.

In the tradition of not putting all its eggs in one basket, the United States was ready at the end of January 1958 with another small satellite from a different (army) program. *Explorer 1* was successfully put into orbit, and the scientific apparatus it carried—all two pounds of it—reported the existence of an intense belt of radiation around the earth at an altitude of 594 miles, named the Van Allen belt after its discoverer. This was an important basic science discovery because it showed the existence of a region surrounding the earth that contained electrically charged particles trapped in the external magnetic field of the planet. These particles had saturated the radiation counters in *Explorer 1* and were recorded by the scientific ground crew that monitored the launch. By mid-March 1958, *Vanguard 1* joined *Explorer 1* in orbit, and U.S. confidence in its fledgling space program was beginning to rebound. However, the intense competition among the armed services and the National Advisory Committee on Aeronautics, each determined to assume responsibility for a viable space program, did not recommend a group effort for the U.S. space program of the future. Nevertheless, there was widespread public and federal consensus that a single, augmented space program was essential; the only question was: who would run it? By March 1958 President Eisenhower and his newly appointed science adviser, James R. Killian, formed the administration's position, which was in general agreement with the finding of Lyndon Johnson's Senate subcommittee on space, namely that the U.S. space agency should be a civilian agency, with the National Advisory Committee on Aeronautics (NACA) as its nucleus. This committee had existed since 1915 and had a large, experienced staff, well-equipped laboratories, and a well-earned record of research performance in aircraft flight. It would provide the core of a strong space program, and at the same time the peaceful, research-oriented nature of the program would mostly avoid projecting the tension of the cold war into outer space, an important consideration to the administration and Congress at the time.

In April 1958 a bill to establish the National Aeronautics and Space Administration (NASA) was submitted to Congress. Both houses had already formed select space committees, and on July 29 President Eisenhower signed into law the National Aeronautics and Space Act. In August T. Keith Glennan, president of the Case Institute of Technology and former commissioner of the AEC, was nominated and confirmed as NASA's first administrator.

By October 1 Glennan was able to announce that the NACA assets—eight thousand people, three laboratories, and two experimental flight stations, with facilities valued at $300 million and an annual budget of $100 million—had been transferred to NASA. The forty-three-year old NACA had come to an end. At that time, the president also transferred to NASA Project *Vanguard*, its 150-person staff, and the lunar probe and rocket engine programs from the army and air force.

The first two years of NASA were a period of organization, innovation, and activity. Design and operations groups had to be formed. For example, Project *Mercury*, the United States' first manned spaceflight program, began in 1958 and needed a worldwide satellite tracking and data acquisition network. And powerful new launch capabilities were urgently required to supplement the existing *Redstone*, *Thor*, and *Atlas* launch vehicles. Planning began for *Scout*, a low-budget booster that would put small payloads into orbit; *Centaur*, a liquid-hydrogen upper-stage booster that promised higher thrust for bigger payloads; *Saturn*, which, with proper upper stages, would put more than 46,000 pounds in Earth orbit and be ready by 1963; and *Nova*, several times the size of *Saturn*, for manned lunar flights in the 1970s. *Centaur* and *Saturn* were already in progress in the Department of Defense space program.

Other major space research programs and the facilities and staff that went with them also moved to NASA. The government owned the Caltech Jet Propulsion Laboratory (JPL) then under contract to the army and an integral part of the army's rocket program. An installation at Huntsville, Alabama, was the center of the army's military rocket program and housed the powerful *Saturn* booster project and a four-thousand-person Development Operations Division headed by the controversial Wernher von Braun, a dynamic German rocket expert from the flying bomb (V-1, V-2) era of WWII. Von Braun was a gifted engineer whose personable qualities also enabled him to make the transition successfully from Nazi Germany to the United States deep in the cold war. Over the strenuous objections of the army, both JPL and Huntsville were transferred to NASA.

The early launch record with existing boosters, however, was not satisfactory. By the end of 1959 more than two-thirds of the thirty-seven

FIGURE 4.1. *Top*: A NACA TEAM CONDUCTS RESEARCH USING THE VARIABLE DENSITY TUNNEL IN 1929.

Source: R. E. Bilstein, *Orders of Magnitude: A History of the NACA and NASA, 1915–1990*, NASA SP-4406 (Washington, D.C.: NASA, 1990), p. 8.

Bottom: A Vought 030 set up for tests using the full-scale wind tunnel at Langley Field, completed in 1931.

Source: R. E. Bilstein, *Orders of Magnitude: A History of the NACA and NASA, 1915–1990*, NASA SP-4406 (Washington, D.C.: NASA, 1990), p. 17.

launches failed to attain orbit. The space program appeared to be a questionable venture, especially in view of its extremely large cost. This concern remained as the Eisenhower administration drew to a close. After the 1960 election, the new Kennedy administration criticized the program's lack of progress and scrutinized its ballooning budget. A committee chaired by the new science adviser to the president, Jerome Wiesner, professed skepticism about NASA's future.

Once again, the USSR resolved U.S. doubts. In April 1961 Soviet Cosmonaut Yuri Gagarin rode *Vostok 1* into orbit, 24,800 miles around the earth, reentered the atmosphere, and landed safely. The question of money and priority for NASA was answered: Congress and the new president pushed ahead with a previously formed ten-year plan for NASA. But to what end? To formulate an answer, Kennedy chose a new administrator of NASA, James E. Webb, owing in part to his reputation for managing large projects in industry and for directing the Bureau of the Budget in the Truman administration. Webb named Hugh L. Dryden, who had been director of NACA and deputy to Glennan, as his technical deputy; the associate administrator and general manager of NASA was Robert C. Seamans Jr., another experienced NASA veteran.

The space program needed a goal that would require more of it than the ability to launch satellites efficiently and catch up with the Russians by putting a man in Earth orbit. President Kennedy was equally in need of a goal to divert the American public from its preoccupation with the cold war. The new NASA administration proposed putting a man on the moon and returning him safely to Earth as the goal of the U.S. space program. It was a risky but ambitious response to the Soviets, and it appealed to Kennedy as a new initiative for the nation. He proposed it to Congress on May 25, 1961, saying:

> Now it is time to take longer strides—time for a great new American enterprise—time for this nation to take a clearly leading role in space achievement, which in many ways may hold the key to our future on earth.
>
> ... I believe that this nation should commit itself to achieving the goal, before this decade is out, of landing a man on the moon and returning him safely to earth. No single space project in this period will be more impressive to mankind, or more important for the long-range exploration of space; and none will be so difficult or expensive to accomplish.[1]

Project *Apollo* changed NASA irrevocably. Previously, NASA had been a multifaceted agency, cautious and expensive, aiming to satisfy several important but loosely connected goals. After Kennedy pledged to put a man on the moon in the 1960s, however, NASA's planning grew more single-minded and more risk and expense tolerant. No new scientific or technological breakthroughs were necessary, but the size and power of the lunar launch vehicles and spacecraft presented problems that were an order of magnitude greater than any NASA had ever encountered. The 1.5-million-pound-thrust boosters needed to launch *Apollo* spacecraft demanded a new logistics system that would take components from the design stage to the launch site with new efficiency and speed. Huge new test stands and launch complexes intended to handle the multistory boosters and spacecraft were too large to be moved by rail or truck. The only option was to employ ship transportation. This required that new NASA centers be located near navigable bodies of water. The Michaud Ordnance Plant outside New Orleans, where the ten-meter diameter *Saturn V* first stage would be fabricated, was on the Mississippi River; the Mississippi Test Facility, with its giant stands for static firing tests of booster stages, was just off the Gulf of Mexico. And the major Launch Operations Center at Cape Canaveral, Florida, which required the purchase of 110,000 acres of Merritt Island, had both water access and sufficient isolation for safety. The resources of the Army Corps of Engineers were called on to undertake construction and provide facilities, just as it had done for the Manhattan Project.

At the heart of this planning was the issue of precisely how to put men on the moon and how to get them off. Initially, it was thought that a big enough booster would allow a direct flight to the moon and the landing of a large vehicle, some part of which—containing moderate-power boosters and a modest-size spacecraft—would return directly to Earth. Technical problems caused this idea to be discarded early in the planning stage, however. Instead, a complex but feasible plan was adopted, one that would require for the first time a lunar-orbit rendezvous of spaceships. In this plan, the launch of a massive mother spacecraft into Earth orbit would be followed by the dispatch from the mother craft of a set of nested spacecraft into a moon orbit. That set in turn would send a smaller craft to land on the lunar surface, let the astronauts explore, and then rejoin the lunar-orbiting spacecraft for the return trip to Earth. This required putting payloads of 300,000 pounds in Earth orbit and 100,000 pounds in lunar orbit. These payloads were demanding enough, but could a rendezvous of spacecraft be made routinely, and, if so, could they dock without a calamity? Project

Gemini was initiated in January 1962 to answer these and similar questions and bridge the conceptual and hardware gap between Project *Mercury* and Project *Apollo*.

Both *Mercury* and *Gemini* carried out a long series of spaceflights of increasing technical proficiency that provided the data on which to base *Apollo*. Doubts and reservations about the ultimate success of the *Apollo* mission remained but decreased as the accomplishments of *Mercury* and *Gemini* increased. The speech that President Kennedy was on his way to deliver in Dallas on the terrible day of his assassination was to begin thus: "This [Apollo] effort is expensive—but it pays its own way for freedom and for America. There is no longer any doubt about the strength and skill of American science, American industry, American education, and the American free enterprise system." Indeed, it was an expensive effort. Originally anticipated to be a ten-year program costing an average of $1.5 billion per year, the NASA budget went from $967 million in 1961, to $1.33 billion in 1962, $3.67 billion in 1963, and $5.1 billion in 1964 and several years thereafter. By that time, NASA directly employed thirty-six thousand people and its contractor and university forces increased that number to four hundred thousand.

As the magnitude of the *Apollo* project was realized, a charge was leveled against NASA that it was solely directed to reaching the moon and ignored the more immediate problems on Earth. This charge was not completely justified. Given the priority of *Apollo*, NASA nevertheless launched the first active communication satellite for the American Telephone and Telegraph Company (AT&T) in 1962. Within a decade, communication and weather

FIGURE 4.2. *Opposite page top*: NASA's seven original astronauts were all experienced test pilots. Posed in front of a Convair F-106, they are (*left to right*) Scott Carpenter, Gordon Cooper, John Glenn, Virgil Grissom, Walter Schirra, Alan Shepard, and Donald Slayton.

Source: R. E. Bilstein, *Orders of Magnitude: A History of the NACA and NASA, 1915–1990*, NASA SP-4406 (Washington, D.C.: NASA, 1990), p. 58.

Opposite page bottom: Kennedy Space Center as it appeared in the mid-1960s. The 350-foot-tall *Saturn V* launch vehicle emerged from the cavernous Vehicle Assembly Building aboard its crawler and began its stately processional to the launch complex three miles away.

Source: R. E. Bilstein, *Orders of Magnitude: A History of the NACA and NASA, 1915–1990*, NASA SP-4406 (Washington, D.C.: NASA, 1990), p. 70.

satellites would be essential in daily life on Earth. Moreover, NASA's development of microelectronics for monitoring the health of astronauts soon gave rise to the everyday use of advanced techniques for patients in hospitals throughout the nation.

Apollo needed more than powerful boosters and giant launch sites, however. It also needed scientifically trained people and a lot of them. As the magnitude of the brain drain grew, NASA was accused of monopolizing valuable resources, chief among them scientific manpower. This was a valid accusation, and NASA felt compelled to meet it with a program of support for science and scientists in universities that followed the precedents set earlier within the science establishment. Beginning in 1962 NASA paid for the graduate education of five thousand scientists at a cost of $100 million and spent another $82 million on university campuses. This program ended in 1970, but contracts and grants for university research rose from $21 million in 1962 to $101 million in 1968 as NASA broadened its effort to include universities as junior partners in the space enterprise.

In short, the infant civilian space agency, NASA, despite the urgent demands of its lunar mission, managed to avoid public contention with the armed services as it competed for resources and planned for its future work in space. It accomplished this by supporting university science and engineering departments, as had U.S. science agencies before it. By 1965 the successes of Projects *Mercury* and *Gemini* provided good reasons to believe that the rapidly maturing Project *Apollo* would be successful also.

The Atomic Energy Commission addressed new challenges of "Atoms for Peace"

The three major problems that the AEC faced when it began business in 1947 were the deterioration of the science capabilities of its laboratories, particularly Los Alamos; whether or not to proceed to develop the hydrogen bomb; and at what rate to move toward use of the atom for peaceful purposes, as in the generation of electric power on a commercial scale. The commission had successfully solved the first two problems by 1955, but the issue of control of commercial power generation, which had been a subject of intense debate during deliberation of the Atomic Energy Act in 1946, remained to be addressed.

As early as mid-1953 the commission formulated a plan to bring nuclear reactor technology into the marketplace. The plan involved acquiring significantly more technical information than was available at the time and modification of the secrecy provisions of the Atomic Energy Act. Qualified information would then be made public. The power utilities would be able to evaluate the magnitude of the technical problems and the investment required of them. To obtain the additional information, the commission recognized that basic research on materials and power reactor types would have to be conducted separately in its own laboratories. It seemed likely that the commission would also be forced to build a nuclear power reactor to provide data on the performance and cost of a full-scale system. With that experience in hand, the reasoning went, it would be possible to assist industry in designing and constructing economically viable full-scale power reactors.

This plan was realistic and would have been possible given the new relaxed classification and security rules in the Atomic Energy Act of 1954. But the role of the AEC and its laboratories in transferring nuclear technology from laboratory to power plant was seen as unacceptable federal interference in the private sector and too slow and cumbersome compared to previous technology transfers. Critics referred to radio broadcasting and commercial air travel, neither of which had depended for growth on government participation and promotion to the extent that nuclear technology promised under the commission's plan. The Joint Committee on Atomic Energy pointed out that radio broadcasting and air travel had been accomplished piecemeal with modest investments by many entrepreneurs moving quickly and with minimal government impedance. But those free-wheeling procedures, the commission argued, were not possible for the development of nuclear power, as it would require huge investments in research before a full-scale plant could be designed, much less constructed and operated. Moreover, the AEC cautioned, there were extremely dangerous aspects to nuclear power that demanded government participation and supervision. How then, was the transfer of nuclear technology to be done safely and fairly and within the constraints of the free enterprise system?

This difficulty did not come as a surprise to the commission or the joint committee. It had arisen during the hearings on the Atomic Energy Act of 1946, when the observation was made that Congress was about to establish an administrative agency vested with unprecedented sweeping authority and entrusted with portentous responsibilities. The act would create a government monopoly of the sources of atomic energy and make that field an island of socialism in the midst of a free enterprise economy. Nevertheless, the situation in 1954 contained no acceptable solution to the problem. In the opinion of the AEC's director of reactor development, Lawrence R. Hafstad, reactor technology was not then well enough developed to allow the construction of full-scale power reactors for commercial use, and there was a limit to what could be learned from paper studies. A veteran scientist and science administrator, Hafstad had directed the Johns Hopkins Applied Physics Laboratory, which had produced the proximity fuze, and had later served with Vannevar Bush on the research and development board of the Department of Defense before taking leadership of the commission's reactor division in 1949. This experience gave weight to his opinion in both the commission and the joint committee.

Hafstad was likewise in the difficult position of steering a course between those who advocated a government-dominated reactor program—concentrating on military projects such as ship propulsion for immediate results—and those who urged an accelerated civilian power program, relying heavily on private industry for development, which would have been unrealistic and dangerous. But neither the AEC nor Hafstad was entirely free to pursue the course of their choice. The commission's reactor development program was heavily committed to reactors for propulsion of ships and planes that preempted available funds and manpower. The chief of the naval reactor program was Captain Hyman G. Rickover, who previously had established himself, with remarkable single-minded determination, as head of the navy's reactor program and was the top official in charge of commission reactor laboratories before Hafstad joined the AEC. With the concurrence of the commission and the joint committee, Rickover gained control of three reactor research facilities: the Argonne National Laboratory, near Chicago; Westinghouse's Bettis Laboratory, outside of Pittsburgh, Pennsylvania; and General Electric's Knolls Atomic Power Laboratory, near Schenectady, New York. All were dedicated to producing a seaworthy nuclear submarine by January 1955. With that control, he was able to supervise the decisions of his contractors and focus on technical obstacles that threatened his timetable. He bypassed small reactor experiments and set out simultaneously to build prototypes of two propulsion systems, one at Bettis and one at Knolls. By mid-1953 he had a prototype operating at the commission's Idaho Test Site, generating enough power to carry a submarine across the Atlantic. With added hard work, a nuclear power plant was made ready for the submarine *Nautilus* by late 1954. This achievement, more than any other single event, convinced the inexperienced, overly optimistic joint committee that nuclear power was a reality and ready to be taken over by private industry.

FIGURE 4.3. *Top*: Planning the development of nuclear-powered ships. Captain Rickover with General Electric and government officials in Schenectady, summer 1946. *Left to right*: C. Guy Suits, John J. Rigley, Hyman G. Rickover, Leonard E. Johnson, and Harry A. Winne.

Source: R. G. Hewlett and Francis Duncan, *Atomic Shield: A History of the U.S. Atomic Energy Commission*, vol. 2, *1947/1952* (Washington, D.C.: U.S. Atomic Energy Commission, 1972), p. 142.

Bottom: Submarine thermal reactor, mark I, Idaho. The land-based prototype as it appeared in 1954. The reactor is located within the portion of submarine hull surrounded by water.

Source: R. G. Hewlett and Francis Duncan, *Atomic Shield: A History of the U.S. Atomic Energy Commission*, vol. 2, *1947/1952* (Washington, D.C.: U.S. Atomic Energy Commission, 1972), facing p. 142.

Power reactors for industry presented different problems, however. Experience with smaller, limited power reactors did not provide enough information to allow extrapolation to a pilot plant and certainly not to proven full-scale reactor technology. There were many examples of this. A plutonium breeder reactor, developed at Argonne as a test of principle, was built at the Idaho Test Site, using liquid metal coolant, and although it showed technical progress, its methods were difficult to use. It was not likely to be a model for a commercial power reactor and was not funded further. A homogeneous reactor also showed initial promise because it had the advantage of eliminating expensive component fabrication. By placing a single fluid mixture of fissionable material, moderator, and coolant in a properly configured tank, a critical mass was produced, and consequently a chain reaction. The experiment produced a few watts of electric power and illustrated that a homogeneous reactor was possible, but it did not indicate how to overcome the serious problems of handling the highly radioactive and corrosive fluid continuously on a large scale. The most promising reactor types were those using water as moderator and coolant. For his submarines, Rickover went to a reactor using water under pressure, which prevented boiling and local power surges thought to be dangerous. Later, boiling water reactors were shown to have higher thermal efficiency than pressured water types, and it was further discovered that local boiling-induced power surges did not give rise to uncontrolled instabilities but would shut down a system if boiling became too severe. Still, the bottom line was that none of these designs was then a practical or economical model for a full-scale power source. Rickover's power plant for the *Nautilus* was the model the joint committee pointed to, but its capacity was limited by the needs of its task, and it was not economical or intended to be economical; it was not the immediate answer to the quest for a peacetime, commercial nuclear power reactor.

The nuclear powered submarine would, however, completely change undersea warfare and influence U.S. military strategy for containment of cold war Soviet expansionism. That strategy focused on hydrogen weapons as a deterrent of unprovoked Soviet military action. The idea was to put the USSR on notice that such action, even if not overtly directed at the United States, would draw U.S retaliation by hydrogen bombs aimed at the USSR. To make the warning credible, the United States developed a multipronged system to deliver bombs throughout the USSR, from the air by the Strategic Air Command, from the land by intercontinental ballistic missiles located in the United States, and from the sea by the nuclear-propelled submarine fleet.

It turned out that the commission had no solution to the problem of devising a commercial nuclear power reactor. Despite seminars and meetings intended to encourage participation and investment by large utility groups, a stalemate with industry persisted. The high cost of developing nuclear power and its difficult technical problems frightened utility executives away from risk taking. This attitude was intensified—not eased—by the successful operation of a commission-financed pressurized-water reactor at Shippingport, Pennsylvania, in December 1957. That first trouble-free, complete full-scale nuclear power plant in the nation reached its full net power rating of sixty megawatts of electricity in the same month that it was commissioned. It had been designed and constructed by Rickover and the staff at Bettis following the engineering practices developed for the Nautilus project, but it was not simply a scaled-up version of that plant. The planned performance of its components had demanded new levels of design engineering and fabrication. The reactor core, for instance, consisted of almost one hundred thousand fuel elements, each encased in a little-known element, zirconium. The decision to use uranium oxide in the fuel elements in slightly enriched rather than fully enriched form had been made after many months of research and testing that produced fundamental engineering data for the future. Shippingport showed the intensive nature and extended scope of the research required for the development of nuclear power reactors at the time.

The total cost of the Shippingport reactor was estimated at sixty-four mills (6.4 cents) per kilowatt of capacity, compared to six mills per kilowatt for existing conventional power plants. Utility executives found this unacceptable, even discounting that Shippingport had broken completely new ground and incurred heavy expenditures to complete the plant by a set deadline. Furthermore, the criticism went, the plant proved nothing because it had not been built by private industry to commercial specifications. The significance of Shippingport went largely unappreciated, as did its public training courses in reactor safety and operation organized during the next six years by the Duquesne Light and Power Company. These courses taught more than one hundred engineers and technicians from the United States and ten other countries the rudiments of reactor technology.

Despite strong pressure from the joint committee and a Democratic Congress, the AEC was restrained by its chairman, Lewis Strauss, and the Eisenhower administration from going beyond the Shippingport reactor. Higher-capacity power reactors were seen as a government-financed program that

FIGURE 4.4. The Shippingport Atomic Power Station in Shippingport, Pennsylvania, was constructed during the mid-1950s to develop and demonstrate pressurized water reactor technology and to generate electricity. The reactor was fueled with three different types of cores, the last being a light water breeder core. The station was shut down in 1982, after completion of the breeder demonstration program. Plant disassembly demonstrated the safe and economical decommissioning of a commercial power reactor.

Source: T. R. Fehner and Jack M. Holl, *Department of Energy, 1977–1994: A Summary History* (Oak Ridge, Tenn.: Office of Scientific and Technical Information, 1995), p. 14.

would build nuclear power plants and establish a government monopoly in nuclear power, similar to the Tennessee Valley Authority's monopoly in conventional power. That prospect was anathema to the Republican Party. As a result, when Strauss retired as chairman in June 30, 1958, the commission had not been able either to formulate nuclear power policy or to promote the development of nuclear power. That was the situation that greeted the new chairman of the AEC, John A. McCone, during his exploratory tour of Shippingport, Bettis, Knolls, and the Idaho Test Station. McCone was a construction engineer who had become president and director of the Bechtel-McCone Corporation when it was organized in 1937. During WWII, he had been executive vice-president of the Consolidated Steel Corporation and president of the California Shipbuilding Corporation. In addition to his business activities, McCone served as special deputy to Secretary of Defense James Forrestal and as undersecretary of the air force in charge of procurement. Eisenhower had previously shown confidence in him.

The director of reactor development, W. Kenneth Davis, who succeeded Hafstad, resigned when Strauss did, and McCone turned to Rickover to give him his first glimpse of the commission's reactor program. McCone was impressed by Shippingport and understood fully that it was not a power plant but a laboratory tool. He did not dismiss it, as some industry leaders had previously. He was also troubled that company engineers at Bettis and Knolls were proceeding to install in commercial reactors fuel assemblies with cheaper and possibly less dependable materials than Rickover had specified in the navy projects. In his personal notes after the trip, McCone wrote: "As a result of these discussions, I am convinced that our reactor division must make the most penetrating study of how the commercial people intend to answer their core design and construction problems. It seems to me that it will be the center of our problem both from the standpoint of economics and ultimate success and safety." McCone intended, he said, to take a constructive approach to nuclear power but not to proceed with "anything which is unsound."[2]

As the fourth chairman of the AEC assumed office, the commission launched into another intensive research and development program, this one to provide critical oversight of the transition from conventional to nuclear power plants. It would not be solely a matter of regulation and equitable treatment of utilities and consumers alike but also involve acquiring and disseminating technical information about radioactivity and the dangers of the enormous energy residing in nuclear power reactors. Much of the work would be done in its own laboratories, but much—the actual construction of power plants—by private industry. And the commission needed to establish a harmonious working relationship with that sector.

The development of nuclear power in the United States was not, however, the only R&D program required of the AEC at the time. Two years after taking office, President Eisenhower had given a stirring speech to the United Nations General Assembly in which he announced the Atoms for Peace program and pledged that the United States would "devote its entire heart and mind to find a way by which the miraculous inventiveness of man shall not be dedicated to his death, but consecrated to his life."[3] Specifically, he proposed establishing an international atomic energy agency and expressed the willingness to share peaceful U.S. atomic energy technology with an international body. Implicit in his promise was renewed vigor in the search for international control and inspection of atomic weapons. There was enthusiasm for the general features and spirit of the program, but the question of control and inspection of fissionable materials and weapons raised by the president's speech caused grave concern in the commission and the joint committee. Soon after Eisenhower's speech, for example, people realized that the abundance of uranium then available in the world made it very difficult, if not impossible, to assure that all fissionable material was declared and accounted for, short of a system of continuous and unimpeded inspection in all countries.

More touchy and immediate were the technical problems created by the worldwide demand to ban atomic weapons testing and to promote disarmament. These involved complex issues that caused nations to align themselves differently from what might have been expected. Predictably, the Soviets emphasized the need for political agreement on a test ban, leaving the method of verification for a later time. They did not take seriously the details of inspection and control necessary to a disarmament agreement since they had no thought of allowing foreign oversight or inspectors into their country. The British objected to a test ban unless the United States was willing to share the nuclear information that had been acquired previously in many tests, but the USA was unable to do this without amending certain restrictive provisions of the Atomic Energy Act of 1954. Nevertheless, Eisenhower succeeded in convening a conference of experts in Geneva, Switzerland, in July 1958. The stated purpose of the conference was "To Study the Methods of Detecting Violations of a Possible Agreement on the Suspension of Nuclear Tests."[4] The U.S. representatives were James B. Fisk, vice president of Bell Laboratories and a member of the president's science advisory committee, Ernest O. Lawrence of Manhattan Project fame and Robert F. Bacher, former AEC commissioner and member of the science advisory committee. Hans Bethe, professor of physics at Cornell University, and Harold Brown, associate director of the AEC's Livermore laboratory, were assigned to advise the U.S. representatives.

Although the Geneva Conference purported to be a meeting of experts on technical issues, it was really an attempt by the president to begin an international dialogue on nuclear weapons and by the State Department, led by Secretary John Dulles, to press for relief from the acute tension in the world at the time. Moreover, there were strong differences of opinion within the government and among scientists on the benefits and disadvantages to the United States of a test ban. Strauss, by then special assistant on science to Dulles, joined with the commission and the joint committee in opposition to an unlimited test ban. Edward Teller and Willard F. Libby, an AEC commissioner, were also opposed because they wished to perfect small, defensive nuclear weapons to counter the huge Soviet standing army. Others took intermediate positions. Whatever their views, all were agreed that verification, as complete as possible, was an absolute necessity for any long-term ban on testing. And they were equally agreed that it was not then available. There was some likelihood that an agreement to ban tests in the atmosphere could be obtained, because violations of the ban might be detected by aircraft sampling the air currents over the earth. This was the method used by the United States, at Strauss's urging, to detect the explosion of the first USSR nuclear weapon in 1949. But tests conducted underground were another matter. In principle, they could be detected with the same seismic apparatus used to study and monitor earthquakes and similar disturbances. Seismic detection depended on the coupling of the underground explosion to the surrounding Earth that served to transport the shock waves to the seismographs. There was, however, the possibility of decoupling the explosion from its surroundings by first firing a relatively small weapon in a very large underground chamber, thus muffling the seismic waves and confusing the detection systems. This possibility, if it were real, needed to be explored in depth because the only alternative was on-site inspection of areas where test ban violations were suspected, and that was unacceptable to the Russians.

In the meantime, it was argued, various compromises were possible that would involve a test ban of limited duration but under strictly specified supervision. The ultimate decision on how much damage to the U.S. nuclear armament should be tolerated and how much risk of Russian violation of any test ban agreement was acceptable would be made in the White House and not by the commission or its scientists. And that was the way the Geneva Conference turned out. The day after it adjourned, Eisenhower announced that the United States would suspend nuclear weapon testing indefinitely, provided the nuclear powers could establish an effective inspection system and make substantial progress on arms control.

The announcement of a U.S. moratorium on testing caused a flurry of activity among those nations with nuclear capabilities. American, British, and Russian scientists rushed to carry out tests both underground and in the atmosphere before October 31, 1958, the start date of Eisenhower's moratorium. Furthermore, several of the U.S. underground tests raised questions about the data originally used at the Geneva Conference. These and other arguments for and against a test ban sapped the strength of the U.S. movement toward a comprehensive test ban, and the Russian unwillingness to consider on-site inspections without a right to veto any or all of them left the entire question dead in the water.

The breakdown of the test ban negotiations was the result of the mutual mistrust and hostility that continued throughout the cold war. Russian intransigence stemmed from American insistence on monitoring Russian activity, not only with respect to nuclear weapons but in everything concerning its armed forces. And the U.S. spy plane flights over Soviet territory exacerbated Russian paranoia in this regard. American suspicion of Russian belligerence and secrecy had been reinforced by spaceship *Sputnik*. This confirmed the U.S. belief that the Russians were capable of secretly preparing major new military threats and springing them on an unwary world. Determined not to be caught unprepared again, the Eisenhower administration gave scientists and engineers renewed responsibility and influence in the higher councils of government. It also expanded its earlier heavy reliance on the AEC and its laboratories for technical advice on the scientific matters that dominated national security issues. Events obliged the Kennedy administration to maintain that posture. Once again, the AEC remained a research and development agency throughout its second decade, heavily influenced in its practices and outlook by academic and industrial scientists and engineers.

Despite all this, the commission managed to sustain research in a variety of applied and basic problems even while confronting pressing new tasks affecting national prosperity and security. Some of the work was done in its own laboratories and some in industrial and university laboratories, and not all of it was successful. Encouraged by the joint committee and the air force, the commission attempted to develop reactors for military aircraft propulsion, hoping to achieve the same success it had shared with the navy. But the work lacked a promising technical base as well as a convincing purpose. Neither Eisenhower nor his advisers, George B. Kistiakowsky, who succeeded Killian as presidential science adviser, and Herbert F. York, the director of the new office of research and engineering in the Department of Defense, was willing to continue recommending the project to Congress after expenditures of more than $600 million and fifteen years of effort had produced little progress. One of President Kennedy's first decisions in 1961 was to kill the project completely. Greater success along a related line was attained when, stimulated by *Sputnik*, auxiliary power generators that employed small reactors were developed to produce more than ten kilowatts of electricity to be used in space by NASA. This $13 million project was the most successful of all the commission's air and space endeavors.

An especially ambitious and exciting idea for power generation emerged at the time that studies of the hydrogen bomb began. The idea was to develop a power reactor utilizing hydrogen fusion rather than uranium fission. A fusion reactor containing an ionized gas of hydrogen isotopes would rely on an inexhaustible, readily available supply of fuel, and radioactive waste would not be a significant byproduct as it was in a uranium reactor. The drawback was that fusing the hydrogen isotopes and releasing the enormous energy associated with thermonuclear reactions would require the temperature of the gas to be raised to one hundred million degrees, many orders of magnitude higher than any temperature ever achieved in a laboratory. Nevertheless, a fusion reactor promised to be the outstanding accomplishment of the peaceful uses of nuclear energy and the hallmark of the Atoms for Peace program. It received enthusiastic support from the commission and the joint committee.

A laboratory system to confine the enormously hot ionized hydrogen gas in a restricted space by means of strong magnetic fields was designed by Lyman Spitzer Jr., a professor of astronomy and astrophysics at Princeton University, while theoretical studies of a hydrogen bomb were in progress. Confined in that way, the gas could continue to be heated, and fusion reactions would ultimately take place when a high enough temperature was reached. The commission funded Spitzer's studies at the same time that the Los Alamos and Livermore laboratories began theoretical work on other containment systems for the same purpose. But this pioneering work made slow progress until the subject of fusion was declassified and opened to wider participation by other laboratories. Scientists interested in doing long-term basic research to understand the physics principles underlying the behavior of gaseous plasmas began to make progress, and articles on fusion research appeared in open technical journals, including the new specialized journal *Physics of Fluids*. By 1965 perhaps $300 million had been expended by the AEC with no fusion reactor yet in sight. But the goal remained as attractive as ever, and research support was not threatened.

With its strong scientific tradition, the AEC envisioned U.S. preeminence in the basic nuclear sciences as a vital part of the Atoms for Peace program. After WWII, nuclear science had moved along several different paths: nuclear chemistry and nuclear medicine blossomed into full-fledged scientific disciplines, and exploration of the nuclear physics of the entire table of elements enlarged understanding of the nuclear force. But discoveries of unanticipated properties of very high energy cosmic ray particles and experiments carried out at higher energy particle accelerators—the synchrocyclotrons sponsored by the ONR and AEC—opened entirely new areas involving new principles in physics. It was these discoveries that had prompted General Groves's science advisers to recommend the construction of new laboratories and new high-energy accelerators in the waning days of the Manhattan Project. The AEC's General Advisory Committee echoed that sentiment.

As scientists used those accelerators to probe more deeply into the phenomena of the new field of physics—high energy physics, as it was called—new constituents of matter, at least as fundamental as the neutron and proton constituents of atomic nuclei, were exposed for the first time, and the need for still more powerful probes became acute. Once again the General Advisory Committee pressed the case with the AEC and the Joint Committee for more energetic accelerators at new and existing laboratories. The commission chairman, John McCone, was not easily convinced that support of expensive research in a field so new and so removed from the mission of the AEC was justified. It required all the authority the high energy physics community could muster, authority derived from its members' contributions to the national defense during WWII, and the strong support of the president's science adviser to change McCone's mind. But it was done, and the AEC became the godparent of another new branch of science: high energy physics.

In all respects the National Institutes of Health grew at an astonishing rate in the period 1955–65.

The legacy of *Sputnik* provided a bonanza for all U.S. science agencies. Within four years after *Sputnik*, the total NIH budget exploded from $98.5 million to $400 million, and appropriations earmarked for extramural research grew from $55.6 million to $322.6 million in 1960. Universities, medical schools, and hospitals generated a burst of requests for the recently established institutional grants, while the volume of individual research proposals that required separate review increased from 2,750, averaging $12,600 annually, to almost 8,000 requesting an average of $19,500. That programmatic growth had to be matched by administrative growth in the DRG. The formerly compact Division of Research Grants ballooned to a complex hierarchy with five operational branches and a staff of 360 full-time positions, and the number of study (review) sections began a steady growth.

The sudden affluence opened doors of opportunity that had been tightly shut before. Over a three-year period, the DRG was able to dispense $90 million in matching grants for university and hospital construction projects. Moreover, the division was in a position to stimulate the development of new basic science areas necessary for advances in medicine. For example,

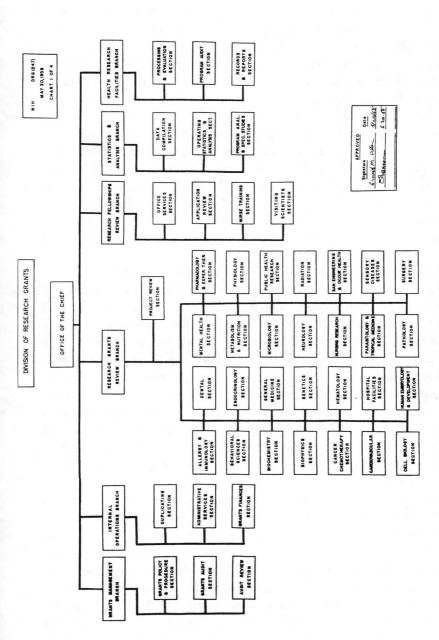

FIGURE 4.5. Table of organization of the Division of Research Grants of the NIH in May 1958.

Source: Richard Mandel, *A Half Century of Peer Review (1946–1996)* (Alexandria, Va.: Division of Research Grants, National Institutes of Health, Logistic Applications, 1996), p. 71.

electron microscopy and X-ray crystallography, both relatively new to medicine at the time, were throwing light on fundamental molecular structure. The Biophysical and Biophysical Chemistry Study Section was able to use a four-year, $600,000 grant to support conferences and lectures in those fields as part of college curricula. Similarly, the Morphology and Genetics Study Section, another area of nontraditional medicine, was enabled to catalyze the field of cell biology by conferring grants on university centers, founding a national journal and a national society, and promoting the idea of a separate institute within the NIH. In 1958 the Cell Biology Study Section became independent, and a dozen working groups of university-based scientists were organized. The Radiation Study Section advanced radioisotope use in diagnostics by awarding grants and organizing conferences on current projects related to radiation effects on health, this at a time when atom bomb testing in the atmosphere was still being conducted and a matter of deep concern throughout the world.

The new ventures were a mixed blessing, however, because they added to the already huge DRG administrative workload. The elaborate review and follow-up procedures carried out for each grant application were overwhelming the study sections and the DRG staff. An ad hoc high-level advisory committee formed to address the situation recommended that an increasing proportion of the NIH research budget go to program and institutional grants as opposed to grants to individual investigators. The committee hoped that this change in emphasis would relieve some of the pressure on the DRG. An early step involved the formation of a new division in the NIH, the General Medical Sciences Division, to focus on extramural basic research. A new branch in the DRG, the Statistics and Analysis Branch, was also created to develop a central source of grant application information by means of automated data processing.

This reorientation established the DRG as a monitor of the NIH research and training programs throughout the nation but did not alleviate the administrative workload. Rapid growth soon caused some units of the DRG to disperse to Silver Spring, Maryland, and soon thereafter the entire extramural organization moved to a site away from the Bethesda, Maryland, center of the NIH. Another effect of the growth was that for the first time the General Accounting Office found that supervision of the granting process was inadequate. Proper control of NIH research funds could no longer be guaranteed. This perception arose despite the fact that the study sections were spending 2.6 days per meeting and processing thirty-one applications per working day and that staff assistance to the study sections was strengthened. As a result, the DRG began to tighten its criteria for grant approval based on excellence rather than growth, both for the science's sake and for its own survival. By 1960 the study sections reported a 43 percent approval rate, down from 65 percent in 1956.

At the end of 1960 Ernest Allen, who had been assistant chief under Van Slyke and then chief of the DRG, moved to the office of the NIH director. He left behind a leadership style devoted to facilitating cooperation and understanding between the government agency and the private sector of the medical community. His tenure at DRG spanned the epochal transition from the period in which grants were awarded largely to individual investigators to the period in which grants were awarded to programs and institutions. By the time Allen left the DRG, nontraditional applications of basic science to medical research and medical practice required evaluation by study sections. Allen was confident that the extramural system with its emphasis on peer review could triple in size in the coming decade, but not if it were driven primarily by individual grants. The budget of $351 million that he shaped for 1960 was divided equally among research grants and training, control, and construction of facilities. The research allocation of $182 million was thought to be sufficient to pay for roughly one-half of the new grant applications that were likely to be recommended favorably.

Allen was succeeded as chief of DRG by his deputy, Dale R. Lindsay, a PHS entomologist from the Malaria Control Program. Lindsay was aware that raising the standards for awards would ease but not solve the problems brought on by the growth rate the division was experiencing. He noted that the extramural grants program had thrived under informal and flexible management dedicated to scientific freedom but that the size and diversity of the program required by emerging biomedical technology demanded more in the way of administrative management. New branches in program review and career development were organized, and new study sections, among them sections in accident prevention research and primate research, were created, indicating the strong trend toward the development of research interests outside clinical medicine. Nevertheless, the division remained seriously understaffed as new divisions with the authority to grant awards were formed in the NIH and the PHS. The chronic understaffing was due to a lack not of money but of experienced administrators and the speed with which medicine was changing among all its disciplines.

The NIH and DRG struggled for a decade to stay abreast of the rush of demands made on them, with only minor hands-on, critical oversight by Congress. That situation changed in early 1962, when a subcommittee of the House Committee on Government Operations, chaired by L. C. Fountain, criticized the NIH for failing to implement adequate fiscal control over grantees as promised. The charges were not without merit. The NIH had been slow to carry out changes in audit analysis and control, largely because the new administrative structures in biomedicine left a shortage of time and people to supervise grantees as carefully as before. The Fountain committee gave a foretaste of the increasing criticism that the NIH and DRG would meet as the size of their budgets increased. Furthermore, the NIH and DRG would clash with the Congress about what determined adequate oversight. There were fourteen thousand research projects to be audited in 1962, and Richard R. Willey, deputy chief of the DRG, made clear its position in a comment after a site visit to the Harvard laboratories of noted cancer researcher Sidney Farber: "Any impression that NIH staff are going to maintain effective day-to-day surveillance over the plans and expenditures of such a grant, I feel, would be illusory. . . . Spending $12 million over seven years on that grant was scientifically justified, . . . but to certify that these funds have not been used for patient care or in 101 other technically inappropriate ways," as the Fountain committee required, was clearly beyond DRG capabilities.[5] Willey contended that the NIH needed to take a new look at the problem of grant management and assign that responsibility to an outside organization. The DRG would then focus on grant review and the formulation and coordination of grant policies and procedures.

The Fountain committee hearings did, however, speed up the evolution of a strengthened partnership between the DRG and the universities. The director of the NIH, James A. Shannon, recognized that a grant recipient institution was in the best position to develop the necessary administrative controls that Congress wanted, and consequently the task of policing grants should be left largely to the institutions themselves. Removal of this burden from the DRG was slow in coming. Meanwhile, the Office of the NIH Director extended its control of programmatic functions, nominally to lighten the load on the DRG. In June 1963 DRG chief Lindsay opted for early retirement. He was succeeded by Eugene A. Confrey, a health administrator with a background in statistics and the humanities. Shannon hoped Confrey would develop a new NIH scientific evaluation capability in the DRG and, with it, a central data system to expedite systematic analysis of scientific accomplishment. These would enable the Office of the NIH Director to evaluate better the progress of NIH extramural investigators and, especially important, to convey that progress to Congress and the administration.

In 1964, in part as a result of the Fountain committee's criticism and in part because the NIH directorate also perceived a lack of control by the DRG, the NIH began to reinvent the DRG, which by then had a staff of 514 to service thirty-two thousand applications for research and training grants and fellowship awards amounting to a budget of $773 million annually. To bring about changes in the DRG, Chief Confrey established new operations offices to handle staff functions, in addition to those in the basic five-branch organization under which the DRG had been doing its business. He also directed some of the management and award duties of the DRG to other institutes and concentrated division activities on initial review of grant applications, based as before on scientific excellence. An external committee appointed by President Kennedy to assess the status of the NIH recommended a strong, centralized NIH administration in which the DRG system would continue as the home ground of the individual investigator. The workload stemming from peer review of those applications would be alleviated by an increasing number of programs managed by the individual institutes, with each program supporting many researchers. That advice was largely negated, however, by President Johnson's campaign in 1964 to turn the NIH away from basic research and toward programs that would concentrate on finding cures for cancer, heart disease, and stroke, by then the leading causes of death by disease in the United States.

The quandary in which the NIH found itself was the product of a cycle that began with the remarkable progress of U.S. biomedical sciences during the period since WWII, progress fueled by NIH encouragement and support. Medical and Ph.D. researchers, in most instances educated and trained at NIH expense, were attracted by NIH success and naturally turned back to the NIH for support in their own research careers. And this cycle, feeding back on itself, renewed itself again and again until after two decades the NIH arrived at an annual budget of $1 billion, which made it the largest federal science agency that provided direct support to faculties in U.S. universities.

The National Science Foundation emerged as a valuable national asset.

The NSF was the only federal science agency that was not part of a larger multifunction agency. It was a study in survival of the idea that there was

POSTWAR DEVELOPMENT OF NIH PROGRAMS

**THE INITIAL SHAPE IS FORMED,
1945-1950**

- PHS Act (1944)
- NIH acquires OSRD contracts
- division of research grants
- first training grants
- categorical institutes-heart, dental, mental
- "omnibus act"-arthritis, neurology, microbiology

**THE FRAMEWORK FOR ACTION
CONSTRUCTED, 1951-1955**

- development of extramural policy and procedures
- shaping of advisory and review framework
- emphasis on basic science
- intramural programs extended-clinical center, biologics standards

**THE YEARS OF GROWTH,
1956-1960**

- the deficit in research support overcome-Folsom policy, Bayne-Jones report, congressional action
- investment in resources-health research facilities act, expansion of training

  problems of institutions and careers-general research support, career support

- international health research act

**THE EMERGING MATURITY,
1961-1965**

- policies and procedures assessed
- program framework rounded out-research facilities and resources, general medical sciences, child health
- developmental research emerging
- new concern with transmission of knowledge and technology

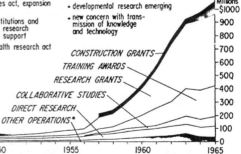

CONSTRUCTION GRANTS

TRAINING AWARDS

RESEARCH GRANTS

COLLABORATIVE STUDIES

DIRECT RESEARCH

OTHER OPERATIONS*

Millions
$1000
900
800
700
600
500
400
300
200
100
0

1945   1950   1955   1960   1965

* State control programs, professional and technical assistance, training, program direction, review and approval, biologics control

value to be gained from the federal funding of science and science education with no strings attached. It tested the dedication of the government to sustaining a federal agency with so narrow and, according to its critics, so impractical a mission.

The director of the NSF, Alan Waterman, had presided at its birth and helped it grow into a vigorous science agency during its first few years. He was beset, however, by the Bureau of the Budget, the watchdog of the executive office, to fulfill two other mandated directives specified by Congress as part of the NSF's mission: he was required to evaluate science research programs undertaken by other federal agencies and to formulate a national policy for the promotion of science research and science education. But recognizing that those were minefields which, once entered by the NSF, might well destroy it, he steadfastly resisted the pressure to take up those no-win challenges. He saw the function of the NSF to be the encouragement and funding of high-quality science research and science education and kept the NSF on that path with as little deviation into national science policy as he could manage. It was soon clear that Waterman's instinct in these matters was correct. National science policy and evaluation of the federal science establishment could only come from the White House, not from the director of a newly established, small agency that could ill afford to make enemies within the government. President Eisenhower's appointment of a science adviser and science advisory committee at the time of *Sputnik* demonstrated that supervision of the government's science agencies belonged in the executive office. The attempt by Congress to assign that function to the NSF was the first of several unsuccessful efforts to place that responsibility within a science agency under congressional control.

The agency held fast to the principle of peer review of research propos-

---

FIGURE 4.6. *Opposite page top*: Presidential visit to the NIH Clinical Center, July 21, 1967. *From left*: Surgeon General William Stewart, President Lyndon B. Johnson, NIH director James Shannon, Assistant Secretary of Health Phillip Lee, and Clinical Center director Jack Masur.

Source: Richard Mandel, *A Half Century of Peer Review (1946–1996)* (Alexandria, Va.: Division of Research Grants, National Institutes of Health, Logistic Applications, 1996), p. 113.

Opposite page bottom: Chart showing the postwar development of NIH programs.

Source: Richard Mandel, *A Half Century of Peer Review (1946–1996)* (Alexandria, Va.: Division of Research Grants, National Institutes of Health, Logistic Applications, 1996), p. 87.

als and study awards and championed award recipients' freedom of choice. Accusations of elitism and failure to distribute funds fairly were bound to occur because the principal business of the NSF was to distribute funds and because the director and policy-setting advisory board of the NSF were politically appointed. To keep contention with Congress to a minimum required courage and wisdom on the part of the NSF's director: the courage to recognize and resist undue interference, different from legitimate oversight, and the wisdom to know when and how to do it. Long after he retired, the second director of the NSF, Leland J. Haworth, told of an incident that illustrated the kind of situation that occasionally arose. This involved a quiet dinner at the home of a senator soon after Haworth assumed the directorship. As the main course was served, the senator remarked to Haworth that the NSF disbursed substantial sums of money each year to individuals and institutions in several states of the union, but his (the senator's) state was not among the most favored. The senator thought Haworth might remedy that situation, which he implied would ensure his future cooperation in matters affecting the NSF when they came before the Senate. According to Haworth, this was all stated gently and tactfully, but Haworth knew that he was being challenged directly and that he had to make a stand, equally gently and tactfully. The response he chose was slowly to push away the plate of untouched food in front of him with the quiet comment that he was unhappily afraid that it was too rich for him. Smiling, the senator pushed the plate equally slowly back to its position before Haworth, with the comment that it was plain and simple fare that would not lead to discomfort. No more was said on the subject, and Haworth—who played no favorites—noted that the senator became his friend and a staunch supporter of the NSF during his directorship. But he often wondered how he and the NSF would have fared if he had failed the test.

The pre-*Sputnik* years of the NSF were marked by modest but adequate budgets that were used in part to introduce new programs and activities that were the seeds of later rapid growth. In the period from 1952 through 1956, science and science education (SSE) amounted to 27.7 percent of the total NSF budget, the remainder going to research and research-related activities (RRA). Most of the RRA expenditures were in the form of awards to individual investigators, but a small fraction went for surveys, travel grants, conferences, and support for data collection and data bases, all of them contributing to the flow of information within the science establishment. During the five-year interval immediately after *Sputnik*, from 1957 to

1961, SSE rose to 39.7 percent of the total budget, and individual investigators continued to dominate research activities.

But a small fraction of research funds began to flow to the support of groups of four or more scientists working in collaboration, for example, on preparation for the International Geophysical Year (IGY). A larger fraction, averaging 19 percent of research funds, went to the support of facilities like the national astronomy observatories and the centers for atmospheric research. The NSF was designated the funding agency and coordinator of U.S. participation in the IGY of 1957–1958, which served as an incentive to support new global atmospheric and oceanographic research and ecological studies. One of the United States' primary interests in the IGY was to bring about an international treaty to preserve the Antarctic for peaceful scientific research. The NSF was made responsible for promoting U.S. interests and instructed to encourage and fund research projects that were best located on that continent.

The visibility of group research left the NSF open to praise for the accomplishments of those efforts and, correspondingly, open to criticism when an ill-conceived project was funded. One such was Project Mohole, said disparagingly by critics to have been planned by a committee. The aim was to dig deep into the earth's crust to explore for the first time the interior of the earth's mantle. Project Mohole never succeeded but was supported at between 5 and 10 percent of the research budget from 1963 to 1969, when it was terminated. It caused the NSF only a moderate headache because Congress was beginning to understand that not all research projects turned out well.

One year after *Sputnik*, the appropriation for the NSF had more than tripled, from $40 million to $134 million, in recognition that the foundation was a vital component of the U.S. response to the questions raised by Soviet advances in nuclear weapons and space technology. Even in the short period between its founding in 1950 and the spaceflight of *Sputnik* in 1957, the NSF became an important supporter of basic research in a variety of scientific fields and a significant partner in science education in schools at all levels.

By 1965 the NSF had survived fifteen years of growth and external pressures, still embracing the values that constituted the reasons for its creation in the first place. It was respected for its manifest determination to hold to those values. More than any other federal science agency, the NSF represented the spirit of science to Congress.

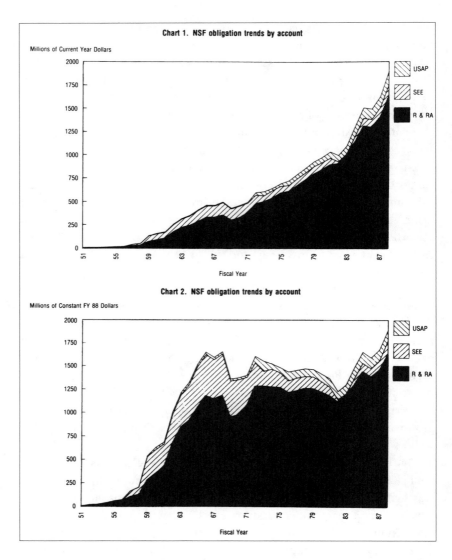

FIGURE 4.7. Charts showing the NSF total budget by account during the years 1951–1987. USAP, U.S. Antarctic Program; SEE, Science and Engineering Education; R & RA, Research and Related Activities. *Top*: current year dollars; *bottom*: constant 1988 dollars.

Source: T. N. Cooley and Deh-I Hsiung, *Funding Trends and Balance of Activities: National Science Foundation, 1951–1988*, NSF 88-3 (Washington, D.C.: National Science Foundation, 1988), p. 4.

CHAPTER 5

End of the Honeymoon: 1965–1975

Entering the decade 1965–1975, the science establishment was an integral component of the research conducted in the United States; by the end of the decade, the establishment was under siege.

By 1965 the science establishment consisted of twenty federal funding agencies, fifty or so private and state universities with large research faculties and graduate student bodies, and several industrial laboratories also pursuing basic research. Laboratory facilities in many universities, neglected during WWII, had been remodeled by federal grants after the war. Facilities that were beyond the financial capability of a single university—laboratories with new, specialized apparatus such as particle accelerators, astronomical observatories with advanced technology telescopes, and hospitals with elaborate diagnostic equipment—were built and subsidized annually to provide U.S. and foreign scientists with modern research equipment. In some instances, the facilities were operated by associations of universities that were responsible for staffing, for equitable use of the facility by qualified scientists throughout the nation, and for the quality of the research produced. The facilities themselves remained the property of the federal government. There were also many federal laboratories, focused on limited areas of research, that were expected to produce important direct applications.

The entire enterprise, excepting only the NIH, was under the benign ministry—as opposed to supervision—of the president's special assistant for science and Technology and the president's Science Advisory Committee (PSAC), both of which had been put in place in 1957, primarily to address science issues related to national security. In practice, the enterprise was self-governing by means of a multitude of loosely connected committees and panels whose recommendations were based mainly on the peer review process.

Under this system, all the federal support of science and technology in the United States—particularly in the universities—flowed through the federal science funding agencies, and it still does today. The variety of agencies gave individual scientists a choice among funding agencies with different interests and different personnel to which they could submit proposals. Equally important, rejection of a proposal for support of a project by one agency was not the end of the world; the proposal could be and usually was resubmitted or submitted elsewhere, often with a successful outcome. In general this loose, unstructured system worked well because the members of each scientific discipline had been educated in pretty much the same way, had similar goals, and were part of the common disciplinary culture.

Within the agencies there were a number of science administrators, many with advanced degrees in their specialties, who found personal fulfillment in encouraging and facilitating research done by others. These administrators helped to set agency policy and were the point of contact with university scientists. They were organizers of the peer review committees and panels, transmitters of recommendations for action within an agency, and ultimately participants in preparing budget requests of Congress. In short, as dedicated public servants of science, they held together the diverse U.S. science establishment that had emerged twenty years after the end of WWII.

The government's commitment to the continuing support of basic science was subjected to its first serious test in the decade 1965–1975, which was not a good one for the nation. Caught up in the many crises of the cold war, the United States defeated the USSR attempt to emplace nuclear warheads in Cuba—the Cuban Missile Crisis—and reacted with increasing irritation and force to the North Vietnamese policy of political murder as a means to conquer South Vietnam. President Johnson greatly enlarged the U.S. presence in Vietnam despite serious misgivings among his own advisers and in the nation as a whole. The military draft was a festering sore on the body of the nation's youth. It divided the country between those who saw service in Vietnam as a moral obligation and those who saw refusal to serve in what they believed to be an illicit war as an equally moral obligation. During much of the decade, the United States was suffocated by the Vietnam quagmire, unwilling either to win or to lose the full-scale war that had developed for fear that exertion of greater force might lead to a larger war.

The war in Vietnam sharply divided most communities in the United States, and the science community was no exception. Members of the PSAC itself were divided on issues arising from the conflict. In response, President Johnson became angry with the divisions within his own White House science structure and was less cooperative with PSAC than Eisenhower and Kennedy had been. Nevertheless, the science adviser and PSAC continued to carry out many studies of scientific, educational, and international concern. Fewer studies relating to national security were undertaken, however.

The succeeding administration under President Nixon did not immediately move to end the war, which gave rise to further unease and impatience among the public at large. The embattled Nixon administration also found it difficult to understand or tolerate criticism, particularly public criticism, of presidential policies by some members of the PSAC. The conduct and official reporting of the war was disturbing to the PSAC, as were the administration's positions on development of an antiballistic missile system and supersonic transport. Disagreements on these issues widened the breach between the president and his advisory committee. Nixon also believed that the Department of Defense had an adequate supply of technical advisory committees and that PSAC should devote most of its attention to nondefense matters, as it had in fact been doing. Nixon's staff was less compromising; they did not understand the work of the PSAC, were especially unhappy with the members who questioned White House policies, and judged the PSAC to be an overall political liability. In January 1973, at the start of Nixon's second term, the position of science adviser to the president was abolished and with it the PSAC and the Office of Science and Technology.

The Johnson and Nixon administrations, deeply preoccupied with the deteriorating situation in Vietnam and the public reaction to it, were seriously upset by the mistrust and disapproval of their policies shown by U.S. university faculties and students. They were particularly ired by the disapproval of American scientists. After all, both administrations contended, scientists were supposed to be a conservative group more likely than most to have the patience to wait for a positive turn of events. Johnson's loss of confidence in the science adviser and the PSAC and their ousting by Nixon were attempts by those administrations to distance themselves from the science community as a whole. Moreover, the war was costing more than either president had anticipated, and fiscal prudence was in order. The budget for the federal science agencies was an expense that could be appreciably reduced, since it did not contribute directly to the prosecution of the war. From a cynical point of view, the reduction would also divert scientists' complaints about the war to complaints about the inadequacy of their funding. Accordingly, the budget of the NSF, always the bellwether agency, sustained a 20 percent decrease in 1969, after peak budgets in 1966–1968. Similarly, the number of permanent, full-time employees of the NIH's Division of Research Grants was reduced steadily by budget cuts in the period from 1969–1974, during which the number of applications submitted for review increased from eight to thirteen thousand.

In the view of the Johnson and Nixon administrations, when scientists took government money to support their research, they became in effect adjuncts of the government, not exactly members but no longer private citizens either. If the Oppenheimer case had not already done so, the hostility of the White House toward its scientific advisers reinforced the idea that elected government officials expected cooperation not criticism from them. And they extended this expectation to the science community as a whole.

Scientists did not immediately react. They would, however, be more wary of affiliations with the government, particularly with the DOD. This wariness coincided with vociferous and unruly demonstrations by students and faculty against classified research on their campuses. In turn, Congress would forbid the DOD from supporting university research unrelated to its military mission. Whether intended or not, the stage was set for the development of an R&D establishment within the DOD essentially independent of and remote from the science establishment outside the government.

The Atoms for Peace program continued to be thwarted by public fear of the atom in any form and by cost competition from conventional energy sources, yet the energy crisis in the early 1970s demanded a unified federal energy policy.

By 1965 atomic energy and atomic weapons had fully emerged from government and military councils into the mainstream of American life. Atomic energy for peaceful purposes and atomic weapons as deterrents of war or as agents for waging war were unrelated applications of the atom. But the public intuitively associated both with the radioactivity they had come to dread after seeing the survivors of Hiroshima and Nagasaki. While most people would pay lip service to the difference between a nuclear reactor and a nuclear bomb, the collective consciousness inextricably connected them to radioactive danger. Neither the AEC nor private industry, both working toward the promotion of nuclear power, recognized the intensity of that aversion to the atom in any form or any application.

It might have been possible to relieve the worst of those fears through strenuous, nationwide educational programs that clarified the difference between bombs and reactors and emphasized the concerns for safety incorporated in reactor designs. The commission was too preoccupied, however, with the technical problems involved in development of power reactors to initiate such a program. And private industry did not then see the problem, much less the solution. The situation actually could have been open to remedy, because the public accepted previously the navy's construction of nuclear submarines in U.S. shipyards, close to population centers, without strong reaction. There was implicit faith—warranted or not—that the navy and the civilian shipbuilders would exercise due care to prevent accidents. That faith might have been carried over to include private nuclear power reactors, but the effort to promote this goal never materialized.

What might have been a redeeming feature of nuclear power—modest capital costs and low operating costs—was not forthcoming at the time of the rush to exploit it. The first power reactor in the nation, at Shippingport, Pennsylvania, in 1957, was described as an economic failure, but it did cause the commission to initiate the Power Demonstration Reactor Program that encouraged industry to design, construct, and invest in other power reactors. By the late 1960s seventy-five nuclear power plants were on order with

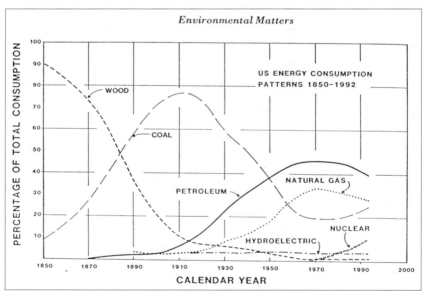

Environmental Matters

US ENERGY CONSUMPTION
PATTERNS 1850-1992

PERCENTAGE OF TOTAL CONSUMPTION

WOOD

COAL

PETROLEUM

NATURAL GAS

NUCLEAR

HYDROELECTRIC

CALENDAR YEAR

a total capacity of forty-five thousand megawatts of output electricity, compared with the fledgling sixty megawatts at Shippingport.

Soon thereafter, the aversion of the public to nuclear power began to be manifested in legal activity. The commission was forced by the courts to consider environmental hazards and other safety concerns in addition to radiation effects and to make substantive changes in its review and licensing procedures. All of this significantly increased the cost of nuclear power plants and decreased utilities' interest in building them.

By then, however, the commission was also worried about a potential shortage of uranium and began intensive study of breeder reactors that would produce more fuel than they would consume. But breeder reactors were also a source of bomb material, plutonium, the material used in the second atomic bomb of WWII. The coupling between the peacetime and wartime uses of atomic energy, which might well have been kept loose without concentration on the breeder reactor, became an intimate connection in the mind of the public and reinforced its fear of any peacetime use of the atom. Moreover, the commission and the utility industry were beset by the growing problem of what to do with the high-level radioactive wastes from nuclear power plants. The only commercial waste-reprocessing plant in the nation had closed down in 1972, and the long, unsuccessful quest for geologic storage sites began about that time. The absence of a waste disposal program in the early 1970s, combined with rising concerns about reactor and environmental safety, dimmed the future of nuclear power in the United States. In other Western countries, however, with fewer fossil fuel reserves, the movement to nuclear power was much stronger and nuclear

FIGURE 5.1. *Top*: Mechanism for removing fuel elements from an experimental breeder reactor, probably the one at Shippingport. This photograph, taken just before full power operation in December 1951, shows the small diameter of the reactor tank in comparison with the large amount of concrete shielding required. During removal a rod had to be shielded and kept in an inert atmosphere at all times.

Source: R. G. Hewlett and Francis Duncan, *Atomic Shield: A History of the U.S. Atomic Energy Commission*, vol. 2, *1947/1952* (Washington, D.C.: U.S. Atomic Energy Commission, 1972), p. 462.

Bottom: U.S. energy consumption patterns, 1850–1992. Note that it has taken roughly sixty years to shift from one dominant fuel to the next.

Source: D. Allan Bromley, *The President's Scientists* (New Haven: Yale University Press, 1994), p. 144. Courtesy of D. Allan Bromley.

power was soon on its way to dominate electric power production in England, France, and Japan.

The nation's ambivalence toward nuclear power and the intense emotions aroused by debate on the subject effectively prevented serious consideration of a broad national energy policy, which was lacking in the United States as late as the 1960s. Historically, Americans relied on private industry to establish production and distribution of the sources of energy, except on occasions when the federal government undertook major energy development projects such as dam building and rural electrification. Even then, federal intervention was regional in nature and restricted to specific technologies. The multiplicity of fuels and resources and their exploitation had caused the government—a large-scale user of energy in many forms—to establish agencies to protect its interests and the interests of individual citizens. The agencies proceeded independently, however, and energy and fuel technologies were not treated in any unified way. For example, the Office of Oil and Gas and the Office of Coal Research, both within the Department of the Interior, acted independently, enforcing conservative practices and fair pricing in the absence of a broader directive. Any such directive would have required some form of explicit national policy and opened the government to the accusation of undue interference in the business of the private sector.

This state of affairs was called into question in 1971 by President Nixon, who observed in a message to Congress that the United States could no longer take its energy supplies for granted. Since 1967, he noted, the United States' rate of energy consumption had outpaced the nation's production of goods and services. Conflicts between energy producers and environmentalists forecast difficult choices that would worsen the energy situation. To address the problem, Nixon asked Congress to establish a department of natural resources to unify energy resource development. That proposal made little headway in Congress or with the public, since both were convinced that the prospect of an energy shortage could not be taken seriously, not at a time when fuel for their cars and homes and industries was ample and seemingly inexhaustible.

The situation changed drastically in October 1973, when war broke out again in the Middle East. The Organization of Petroleum Exporting Countries (OPEC) placed an embargo on crude oil shipped to the United States. A month later, oil supplies were critically low, creating acute shortages of fuel for cars and electric power plants. Nixon's warning had been validated, but a public opinion poll in January 1974 did not confirm that the U.S. public believed the shortages to be real. The responsible parties, they thought, were the oil companies and the federal government, not necessarily the Arab nations.

Nevertheless, the Nixon administration moved rapidly to establish the Federal Energy Administration, a temporary agency intended to meet the energy crisis. The Watergate scandal forced President Nixon to resign in August 1974, but soon after, amid the national turmoil provoked by his resignation, Vice President Gerald R. Ford assumed the presidency and signed the Energy Reorganization Act. It established the Energy Research and Development Administration (ERDA) and the Nuclear Regulatory Commission (NRC) and abolished the Atomic Energy Commission. The NRC was given the authority to license and regulate nuclear power plants, originally functions of the AEC, while the development and production of nuclear power and weapons went to ERDA, which, under the reorganization, also acquired the task of unifying all energy concerns and technologies within the federal government. The energy research and development functions of the Office of Coal Research and the Bureau of Mines also went to ERDA, as did the NSF offices of solar and geothermal development and the Environmental Protection Agency's work on innovative automotive systems.

The Energy Research and Development Administration was activated on January 19, 1975, with Robert C. Seamans Jr. as administrator. Seamans was president of the National Academy of Engineering and a former secretary of the air force. He divided the new agency into traditional units—fossil energy; nuclear energy; and solar, geothermal, and advanced energy systems—and further established units for environment and safety, conservation, and national security (including weapons research and production).

Seamans was required by the Energy Reorganization Act to consult with the secretary of defense to decide whether the nuclear weapons programs should be transferred to the Department of Defense or retained under civilian control in ERDA. One year later, their report to the president recommended that civilian control be continued under ERDA because civilian weapons research laboratories were already uniquely capable to handle such research and development.

The AEC was the first exclusive science and technology agency handling projects of such magnitude that it affected national security directly. Deeply involved in the production, testing, and control of entirely new weapons, all with enormous destructive power, the AEC was fortunately shielded from the outside world and its reactions. Recall that committees were formally prescribed in its enabling legislation: the Joint House and Senate Committee on Atomic Energy, the Military Liaison Committee, and the General Advisory Committee. Taken together, the AEC and its committees were a well-balanced, carefully crafted instrument, fully able to carry out the duties originally assigned by Congress. In the turbulent thirty-year period of unrelenting heavy pressure that followed its creation, it served the nation well.

On the other hand, the circumstances that produced ERDA—the energy crisis and the administrative desire to unify responsibility for national energy resources and operations—ensured that it would be simply another government agency left with the burden of continuously justifying its policies to Congress. There, a wide spectrum of contending opinions on energy policy were represented, often loudly and contentiously. Protection from those contending forces might have come from the inclusion of a standing House/Senate committee on energy to oversee ERDA, but that was never provided. Nor was provision made in the ERDA legislation for scientific and technological support of the new agency by nongovernment committees. The new agency was left to carry its burden alone. The absence of outside support and constructive criticism for the Department of Energy (DOE), which, as planned, would replace ERDA in two years' time, would be an ongoing, serious detriment.

The Daddario-Kennedy amendment brought the National Science Foundation more closely under the wing of Congress.

In 1965, during Leland Haworth's tenure as director of the NSF, a subcommittee of the House Committee on Science and Aeronautics, which had been created in the aftermath of *Sputnik*, began to review the NSF charter. Three years later, at the same time as approval of a $500 million NSF budget, Congress passed a major amendment—the Daddario-Kennedy Amendment—to that charter. The amendment required annual reviews of the NSF's programs by both the House and Senate subcommittees on science. It also required annual authorization for its budget appropriation, replacing the continuing authorization that had been provided in the original NSF act. The amendment also specified that the appointments of the associate director and the four assistant directors be made by the president. Previously, presidential approval had been required only for the director and the Science Board. Besides making the foundation more vulnerable to political interference, the Daddario-Kennedy Amendment sought to bring opera-tion of the foundation more firmly under the control of Congress. Now Congress could explicitly direct the NSF to support social sciences and applied research. Those disciplines, long subjects of contention, were dealt with in the NSF by the experience gained from trial and error. The value of an explicit directive to fund them was at best debatable.

The Daddario-Kennedy amendment significantly changed the NSF, but Congress and the White House were not done yet. The first in a series of events was the selection and subsequent repudiation by the Nixon admin-istration of Franklin Long, a distinguished chemist at Cornell University, to succeed Haworth as director of the NSF. Nixon determined that Long was unacceptable because Long had opposed the administration's antiballistic missile program. Many scientists were already alienated by and highly sus-picious of the ultimate effectiveness of the program, and they deplored its wastefulness. On the occasion of the appointment of only the third direc-tor of the NSF, the original nonpolitical condition for selection was violated.

The directorship was assumed by William D. McElroy, a Johns Hopkins University professor of biochemistry, who set out to move the NSF toward a larger budget and a higher public profile. In 1969, when McElroy took office, the NSF was spending $400 million annually, compared with its $40 million pre-*Sputnik* budget. In that year, however, it experienced the first decrease—of almost 20 percent—in its history, allocated by Congress and the administration. The reasons given for the retrenchment were an over-supply of Ph.D.s and the lack of relevance of the NSF's programs to national problems, particularly the war in Vietnam. To address the budget cut and accomplish his goal of a $1 billion agency, McElroy felt it necessary to work closely with the Office of Management and Budget (OMB), formerly the Bureau of the Budget, in shaping the NSF's programs as well as its expendi-tures. The National Science Board, the NSF's and McElroy's statutory advi-sory board, strongly opposed OMB influence in policy making of the NSF, and, as they feared, it temporarily weakened the agency.

Concern with relevance led McElroy to initiate a program known as Research Applied to National Needs (RANN), which was also a response to the Daddario-Kennedy Amendment. The establishment of RANN was antic-ipated in 1970 by a modest $6 million program with the grandiose name Interdisciplinary Research Relevant to Problems of Our Society (IRRPOS).

Following NSF practice in the basic sciences and engineering, the foundation requested proposals from the scientific community in the areas of environmental quality and urban growth and management. After two years, and without significant accomplishment, IRRPOS expanded into RANN.

The NSF and OMB jointly attempted to design a comprehensive program of technical innovation for a presidential message on science but did not succeed in producing anything persuasive enough for presentation to Congress and the public. However, the attempt did stimulate OMB to promise a $100 million budget increase to NSF in return for its support of a major applied research program focused on national problems. The program was intended primarily as a stimulant to the national economy. A pump-priming program was recommended to most federal agencies by the OMB in the 1972 budget year. The foundation agreed to phase out a major portion of its educational programs in return for the $100 million, part of which would go for applied research. Yet RANN had another budgetary effect on the NSF. The rise of expenditures for RANN coincided with the phaseout and termination of the Institutional Support Program (ISP) of the NSF. From 1960 to 1974 Institutional Support furnished block grants to institutions, primarily universities; funds could be applied to research support, instrumentation, and graduate research facilities. The ISP attempted to improve the science infrastructure of those universities that had suffered in WWII and again in the financially tight years during and after the Korean War. It was a major part of the NSF's growth, accounting for 20 to 25 percent of the research budget during the period from 1960 to 1968. But beginning in 1969, with the NSF's $30 million budget cut, the program decreased rapidly and was essentially supplanted by RANN, even though the content and clients of the two programs were entirely different. By the time RANN was downgraded in 1977 to a small applied research directorate, it had spent almost $500 million. Most of the program was transferred to ERDA and later to the DOE.

These forays by Congress and the OMB into NSF policy and operations were too hit-or-miss in nature to have lasting value or to do lasting damage. Moreover, the steps taken to make the NSF leadership subject to political direction did not accomplish what their proponents intended because committees of Congress were not equipped to exercise detailed day-by-day direction of a science foundation. The NSF was damaged, however, by the increased extent to which appointments of its directors were politicized. An ominous indicator was the resignation of McElroy in 1972, after a tenure of only three years (Waterman had served two full terms, a total of twelve years, and Haworth, six). H. Guyford Stever, who succeeded McElroy, also served a mere three years but that was during the time when the Nixon White House was feuding actively with the science establishment. Stever had been in the OSRD throughout WWII and was chief scientist of the U.S. Air Force afterward, in 1955–1956. He was professor of aeronautics and astronautics at MIT and president of Carnegie-Mellon University from 1965 to 1972. President Nixon assigned him the task of making the science-advising system work while he served simultaneously as science adviser and director of the NSF.

Nevertheless, the core values of the NSF remained intact, and it prospered financially throughout the incursion by RANN, despite the demise of the institutional support program. The growth of RANN coincided with significant increase in individual investigator support. In 1970, when RANN started, individual investigator awards accounted for 50 percent of the NSF budget. By 1978 individual awards reached 66 percent of the total budget, more than twice the dollar amount expended in 1970, and RANN ceased to be supported. The other categories of major expense were group research and facilities, the latter signifying the NSF's support of major experimental facilities, which substituted for institutional awards. The dedication of the NSF to its primary mission—the support of basic science in a variety of fields and mathematics and engineering—gave rise to advances in many of those fields, leading more individuals to compete for grants from the larger NSF budget. The quality of the proposals that were submitted was raised by the competition, and the awards were more widely distributed geographically, a result that Congress originally sought.

Internal and external politics roiled the National Institutes.

The NIH was undergoing its own difficulties, similar to those plaguing the NSF. Two issues vital to the well-being of the NIH arose once more, this time with increased intensity. The first was the issue of peer review, with its implication of elitism. The second concerned the relative value of research initiated by individual investigators as opposed to centrally managed programs with consensus goals. These issues provoked serious discussion within the medical community, but differences therein were exacerbated by congressional interference and especially by the climate of hostility toward science emanating from the executive office. Much of the real internal problem came from the NIH's enormous growth during its short existence. The

DRG organization, which once operated with a personal touch, had become necessarily an impersonal bureaucracy. It still intended to maintain prompt review procedures but was forced to develop cumbersome requirements that alienated many of its applicants, even successful ones. But once funding leveled off, followed by Vietnam War–driven budget cuts, the NIH realized that it had overexpanded to the point that it could not meet its obligations as long as the budgetary crisis remained.

At the same time, the Nixon administration, in what was perhaps a legitimate effort to reduce the federal commitment to biomedical research, waged an unnecessarily vindictive war against the grant system, summarily firing a director of the DRG, suspending training grants, and reducing the number of review committees. Of even greater concern, it attempted to abolish, or at least radically change, the peer review system itself. The OMB in fact called for abolition of the study section function. Only Nixon's resignation in August 1974 thwarted this wrong-headed plan, but by then it had unleashed contending forces within the NIH and DRG and further encouraged discord between the academic science community and the NIH. For example, an NIH planning committee studying administrative reform expressed a preference for review by a new office of extramural services of all submitted proposals, based on what it called "program relevance," prior to the consideration of technical merit. This idea went against the basic tenets of the NIH, and the incoming NIH director, Robert S. Stone, ruled against establishing such an office. On the government side, an assistant secretary in the Department of Health, Education, and Welfare, remarked that the extramural review system was "governed by the public will" and cautioned the NIH against any change in its organization that would reduce the government's involvement in NIH operations.[1]

Nevertheless, the NIH began yet another examination of the DRG to consider limiting its independence in favor of a larger review role for the individual institutes or perhaps to do away with it completely. After expansion in the early 1960s, the DRG was downsized by one-third, and this staff reduction, coupled with a new, artificial ceiling on the allowable number of study sections, worsened the situation. Committees and study teams were formed to debate administrative change and modernization. A complicating factor in these deliberations was the Sunshine Law of 1974, a mandate intended to change the merit review procedures of all government science agencies by opening all meetings to public attendance. The Grants Peer Review Study Team, consisting entirely of NIH officials, was organized early

in 1975 to consider the Sunshine Law, its constraints, and the operating regulations newly issued by the Department of Health, Education, and Welfare. The study team chairperson, Ruth L. Kirschstein, director of one of the
NIH institutes, presented one current of thought of the time by asking the
broader questions: "How can a system, devised in an era of elitism, of
secrecy, and of economic growth . . . be adapted to an era in which stress is
on equal opportunity, openness, and limited availability of funds? . . . If
such a system proves unworkable, what system should be substituted?"[2]

After thirty years of nonstop growth and extraordinary accomplishments, such overstated questions and the self-examination they involved
were unlikely to lead to an improved peer review and award system. Indeed,
the study team brought forth a number of recommendations following

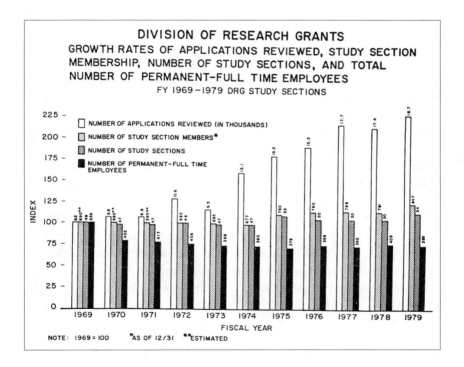

FIGURE 5.2. Ten-year survey of growth rates of the Division of Research Grants in several areas.

Source: Richard Mandel, *A Half Century of Peer Review (1946–1996)* (Alexandria, Va.: Division of
Research Grants, National Institutes of Health, Logistic Applications, 1996), p. 123.

extensive opinion surveys of academic review group members and respondents at public hearings. In the end, the DRG was held up as a model organization, one that expertly carried out the review function. More than 95 percent of those surveyed had rated the DRG review process as "good" or "excellent." Large majorities spoke strongly against congressional demands for public access to review procedures and against charges that the study sections were afflicted with bias, mediocrity, and cronyism. The DRG retained its influence in extramural policy making and review.

It would strain credulity to think that all the real internal problems of NIH had been solved by 1975 and that the organization would function seamlessly thereafter. What had been accomplished was consolidation and survival during a time of fiscal retrenchment and national upheaval. In the face of an administration whose mien was adversarial and dictatorial, the NIH had endured, as had the NSF, with values intact, and the marriage of science and government had been preserved, albeit uneasily.

NASA put a man on the moon and began consideration of its future in the post-Apollo period.

The specter of cost returned to haunt the government's view of the space program. As the decade 1965–1975 began, NASA was moving toward the goal set by President Kennedy, and few in Washington were likely to raise objections to the legacy of the assassinated president. Nevertheless, the NASA program in its entirety was not justifiable in the period of fiscal conservatism brought on by the Vietnam War. The *Gemini* program—to prepare for human experience in space—could not be relinquished if the *Apollo* program to put a man on the moon was to succeed, but a number of peripheral efforts could be delayed or even allowed to lapse in the interest of saving money.

These cutbacks were prudent, dictated by NASA itself for certain peripheral tests and experiments. Neither the White House nor Congress directed the cutbacks. A case in point, involving a reversal of roles, was the Supersonic Transport (SST), a supersonic jet plane intended for commercial airline use. The Kennedy administration endorsed studies of an SST, and NASA, in cooperation with the air force, began a test program that produced a Mach 3 airliner a few years later. That airliner required a titanium airframe instead of the customary aluminum, because titanium could withstand the greater heat generated at the increased speed. The test program produced valuable data on supersonic flight relevant to military aircraft. It also demonstrated that shock waves from sst airliners, even at the slower Mach 2, would prohibit supersonic routes over the continental United States. For reasons of economy, the test program was terminated at the time of the *Apollo* moon expedition.

Later, when the question arose of U.S. entry into the international sst airliner race, nasa had a solid base of experience to contribute to a U.S. effort. Wise counsel from Deputy Administrator Hugh Dryden, however, kept nasa in a supportive but secondary role. He reasoned that the *Apollo* program effectively prevented nasa from sponsoring another expensive program during the same budget years because they would each compete for the same funds. The fate of the sst in the United States, despite its promotion by the Nixon administration, testified to Dryden's wisdom. Furthermore, the tragedy of the three astronauts in the spacecraft originally intended for the *Apollo* mission put in motion a redesign of the spacecraft wiring system and a materials study program that alone cost $50 million. The $450 million and the manpower required by that emergency were available to nasa without competition from another high-priority in-house program.

The space administration approached its budgetary responsibilities in a sensible fashion: the White House and Congress fixed the annual budget under competing pressures from nasa and the omb, but the emphasis and money invested in each component of the space and flight programs were determined almost exclusively by nasa itself. For example, the first discoveries about space from the earliest satellites suggested that it would be profitable to put up bigger, dedicated science satellites. These more complex spacecraft were built and launched in the mid-sixties, but their complexity caused many technical problems, and their results were generally disappointing. By the late sixties, the *Gemini* and *Apollo* programs preempted investment in the larger science satellites. At the same time, the planetary programs were also cut back, and the ambitious *Voyager* program was first curtailed and then dropped in 1968.

A year later, on July 20, 1969, Neil A. Armstrong and Edwin E. ("Buzz") Aldrin landed on the moon, took the first tenuous steps, planted the U.S. flag, and placed scientific apparatus on the lunar surface. After collecting rock samples, as geologists had been doing for centuries on the earth, they returned to the lunar module. The following day they returned to the command module for the flight home and to an ecstatic reception by a relieved and proud nation.

The *Apollo* mission had taken eleven-and-a-half years and had cost $23.5

billion. But by putting men on the moon, NASA acquired the equipment—mental, emotional, and technical—for humans to voyage in space. The world viewed NASA as the premier space agency, and its astronauts were regarded as folk heroes. There might have been cutbacks at the tail end of the *Apollo* program—the last three launches were scrapped, and the NASA budget went from $5 billion in the mid-sixties to $3.7 billion in 1970—but NASA, as seen by Americans, was now a national treasure and a source of glory.

At the same time that *Apollo* was developing, NASA embarked on a program of smaller Earth satellites dedicated to various commercial applica-

FIGURE 5.3. Astronaut Neil A. Armstrong took this photograph of Edwin E. Aldrin Jr., deploying the passive seismic experiments at Tranquility Base on the moon while the ungainly lunar module crouches in the background.

Source: R. E. Bilstein, *Orders of Magnitude: A History of the NACA and NASA, 1915–1990*, NASA SP-4406 (Washington, D.C.: NASA, 1990), p. 90.

tions. These became another major NASA achievement. The potential of communications satellites was recognized almost immediately, and NASA would not need to incur further expense in satellite communications once the research and development was completed. The Kennedy administration fostered the creation of the Communications Satellite Corporation (ComSatCorp), a government-industry collaboration with a strong international component. The corporation was given the authority to invite other nations to share in the investment as well as the service and profits, a feature that provoked a Senate filibuster before the proposal was eventually approved. The first ComSatCorp launch by NASA took place early in 1965. Later, weather and navigational satellites were launched but produced data of limited use, and large flows of cash and effort were not expended on satellite technology at the time. The experiences with these satellites led, however, to a proposal by the Department of Interior for an Earth resources satellite program that in 1972, after the peak of the *Apollo* program, developed into the very successful series of Earth Resources Technology Satellites (ERTS). A ready-made clientele, governmental and commercial, avidly sought information from ERTS—soon renamed the *Landsat* satellites— which yielded real-time data on virtually all of Earth's surface, including spectacular, extremely detailed photographs.

The successes of ComSatCorp and *Landsat* at the time when the *Apollo* mission was coming to an end heralded a new era for NASA, an era in which NASA would serve the human need for increased communication between people and nations, no matter the distance separating them. To a far greater extent than before, NASA would serve also the interests of basic science, physics, astronomy, and biomedicine. One initiative that arose during the 1972 NASA administration of Thomas O. Paine, was the space shuttle, which was proposed to the Space Task Group, a panel convened by President Nixon to advise on prospects for the post-Apollo period. The task group reported favorably on the space shuttle program, and once its cost was halved to $5.25 billion by the new NASA administrator, James C. Fletcher, development of the shuttle was approved.

A new relationship begins.

In the decade 1965–1975 the science establishment entered the real world of a protracted war, severe budget constraints, and a hostile administration in

Washington. However, the well-being of the establishment, nurtured by the federal funding of research, survived even that stressful period. Moreover, the spirit of cooperation between the federal science agencies and university scientists was in the main excellent. The funding level did not permit the growth rate of the previous two decades, but it was adequate to meet the challenges of the science of the time. U.S. science and technology achieved world eminence, despite the national turbulence in those years.

Once Gerald R. Ford acceded as president in 1974, Congress reestablished the Presidential Science Office and the position of science adviser to the president. Congressional approval and support of the science office and science adviser were thought to be necessary, and President Ford signed the bill creating this approval structure. Soon after taking office in 1977, President Carter, with a strong interest in science and engineering, made use of the structure and appointed Frank Press as the new science adviser. This appointment was well received. Under Press's leadership, the science office, newly named the Office of Science and Technology Policy (OSTP), was active once again within the White House.

Nevertheless, the relationship between the government—the White House and Congress—and the science community took on a different character after Nixon and Ford. Both sides recognized a mutual loss of trust and confidence. The easy spirit of consultation and cooperation engendered during WWII and in existence for more than a quarter century had changed. It was not a hostile relationship as it had been under Nixon, but it was stiffer and more formal than it had been before. In 1968 Congress passed the Mansfield Amendment, which forbade the DOD from supporting any research in universities not directly related to the military mission, and though later it would be weakened, this law remained influential in the policies of military and university administrators. Nixon's belief that the Department of Defense should be essentially independent of external science advisers contributed to and confirmed the major expansion of a research and development establishment within the DOD. Finally, Presidents Ford and Carter, while restoring a science advisory structure to the White House, did not reestablish the nonpartisan advisory committee, the PSAC. Those administrations would require science advisers whose loyalty was virtually guaranteed. All these events contributed to a separation—not a divorce but an estrangement—between political Washington and the science establishment.

CHAPTER 6

Estrangement and Reconciliation: 1975–1985

Questions of efficiency and management arise.

In retrospect, it is easy to see that the close rapport between the federal government and the science community would ebb away naturally over the course of time. That rapport emerged from WWII and was sustained by the cold war and the peacetime contributions of science and technology to the quality of American life. But other national cares and worries and a natural tendency to take the science establishment for granted brought about the separation. The reinstatement of a science advisory structure in the executive office of the Ford administration was reassuring, as was President Carter's appointment of a well-respected scientist as his science adviser. Moreover, there was no movement by either executive to make overly large cuts in the science budget despite the need to pay for the Vietnam War and the ongoing cold war. Federally funded science and technology continued to be recognized as a proper responsibility of the government, and the science establishment was regarded as a valuable national asset. Emphasis in Washington in the decade 1975–1985 turned instead to the more pragmatic

issue of how to improve the efficiency and effectiveness of the science establishment through better management.

"Bigger bang for the buck" was by no means a new idea. It had been
raised frequently since the end of WWII as science budgets increased with
such rapidity, but it became more difficult to address as time went on. The
issues of management were twofold. Management that was too loose would
tolerate funds spent poorly on low quality or ill-directed research and
would also limit the support available for superior work. Management that
was too tight would frighten scientists away from more daring and often
more rewarding programs; it could encourage mediocrity even in those
areas of study considered to be most valuable.

Both the executive and legislative branches of the federal government
had searched diligently for a plan that would help them stay informed
about the science agencies and oversee the science research those agencies
funded. For example, one responsibility of President Eisenhower's science
adviser and the PSAC was to report on the status of science both in and out
of the federal government. To strengthen the institutional base for information, Eisenhower also established the Federal Council for Science and
Technology (FCST), whose members were the top-ranking scientific and
technical officers of the federal departments and agencies that housed the
largest research and development programs. The FCST was chaired by the
science adviser and was intended to be the internal equivalent of the PSAC.
Neither the FCST nor the PSAC developed as the desired encyclopedic source
of information or as a critical force for management. Later, the science
adviser to President Kennedy, Jerome Wiesner, became the White House
contact point for almost the entire governmental science apparatus,
although health research remained largely outside his orbit. In the summer
of 1962 the natural expansion of the duties and activities of the science
adviser—particularly in science matters related to national security—led
to the creation of the Office of Science and Technology (OST), the prede-
cessor of the OSTP, within the executive office of the president. It had its
own budget and its own staff and concentrated on the nation's security. It
did not serve as an important presence within the federal science enterprise.

Even if these attempts by the White House had worked better than they
did, Congress would not have been satisfied, because it had only limited
access to the president's science adviser and the OST. At the end of 1963 the
House Subcommittee on Science, Research, and Development began hear-

ings on "Government and Science, to identify problems in the relationship of the government and the science establishment, and to assign priorities for dealing with them." This was one of several attempts by Congress to address the subject, all fruitless. Over the years, Congress made more attempts at information gathering and regulation of the science establishment. Like those before, none succeeded.

Why didn't successive presidents and Congresses bite the bullet and create one single agency that would supervise all science in and out of the government, an agency through which all funding would pass? Actually, it was an old idea, one that arose as early as 1884. The growth of the science establishment after WWII revived the idea and led to a proposal for the creation of a Department of Science and Technology that surfaced in the Senate Committee on Government Operations in 1958. At hearings in 1959 a revised bill to that effect was reported favorably to the Senate, but one finds the same Senate committee considering a substitute bill in mid-1962 for the establishment of a Commission (not a Department) on Science and Technology "to bring about better coordination of the science activities of the federal government." The idea of a commission or a department has been revisited since then, but it never gained wide appeal.

It would be quite wrong to view these bits of history as typical examples of Washington's inability to resolve difficult problems. It was then and is still unclear that a single agency in charge, so to speak, of science and technology would benefit either the nation or the science establishment. It was argued then and is still that such centralization would lead inevitably to a huge bureaucracy, a heavy weight on the free-enterprise spirit so necessary to new ideas and new directions in scientific research and technology. This is the same thought so well articulated in Vannevar Bush's original treatise *Science: The Endless Frontier*. Furthermore, even in the early 1960s it was by no means obvious that the AEC, NIH, NSF, and NASA, apart from other federal agencies, could be fitted into a single department of science in the federal government. True, their scientific efforts were interrelated and reinforced one another, but their diversity—one of the reasons they were separate to begin with—tended to make any merger into a single entity extraordinarily difficult and at best unrealistic. It is not surprising that responsible government has been unable since to find a unifying mechanism to manage the science establishment. Improvements in efficiency and effectiveness have to come piecemeal, in consonance with the piecemeal nature of the science establishment itself.

The Department of Energy was created and given the energy responsi-
bilities of all other federal agencies.

Barely two years after the termination of the Atomic Energy Commission,
President Carter signed into law a bill creating the Department of Energy
(DOE) to replace ERDA. The major provision of that law called for the functions of ERDA to be transferred to the new department, along with those of
the Federal Energy Administration, the Federal Power Commission, and a
number of generally similar functions in the Interior and Commerce
Departments, as well as in Housing and Urban Development and the Inter-
state Commerce Commission. Responsibility for the naval petroleum
reserves of the Department of Defense (DOD) was also placed in the DOE.

The first secretary of energy, James Schlesinger, attempted to organize
and structure the DOE to fit the national energy policy of the Carter administration. The department would be led by the secretary, a deputy secretary,
and an undersecretary. Energy technologies would be grouped under several assistant secretaries "according to their evolution from research and
development through application and commercialization."[1] Basic research
was placed in the Office of Energy Research. Individual research and development projects in solar, geothermal, fossil, and nuclear energy were placed
under the assistant secretary for energy technology. After scientific and
technical feasibility were determined, projects would be transferred to the
appropriate assistant secretary for resource applications or for conservation
and solar applications, both of whom had specialized expertise in commercialization and energy markets. The assistant secretary for environment
would assure that all departmental programs were consistent with environmental and safety laws, regulations, and policies. The assistant secretary for
defense programs would inherit responsibility for the nuclear weapons programs.

To allow for the continuity of programs and functions from its predecessors, all activities of the Federal Energy Administration and ERDA without exception were distributed throughout the DOE. In addition, the Federal
Energy Regulatory Commission (FERC) was established as an independent
agency within the DOE. This five-member commission was made responsible for the licensing and regulation of hydroelectric power projects, regulation of electric utilities, transmission and sale of electric power, transportation and sale of natural gas, and the operation of natural gas and oil

pipelines. Regulatory programs not included in FERC were placed under the Economic Regulatory Administration (ERA), one of two administrations created in the department. The ERA took on oil pricing, allocation, and import programs, most of which had been established during the energy crisis of 1973–1974. A second administration within the DOE was the Energy Information Administration, which consolidated the government's many diverse energy data systems to provide comprehensive data and analysis for the president, the Congress, and the DOE.

At the same time that the DOE developed into an enormous bureaucracy with about twenty thousand employees and an annual budget of $10.4 billion, it was left without the committee structure that so ably supported the AEC. The Joint House-Senate Committee on Atomic Energy was abolished, and its responsibilities assigned to several committees in each chamber. The representation in Congress that a joint committee would have provided was not deemed necessary, although the multiplicity of functions and problems that the DOE had inherited were certain to make that representation imperative. Perhaps the idea of direct congressional oversight was seen as inappropriate for a federal department with a secretary sitting in the president's cabinet at its head. Whatever the reason, the DOE, a vastly expanded version of the AEC, was left to shift for itself as far as Congress was concerned. Other committees were also dismantled: the Military Liaison Committee and the General Advisory Committee, both of which served as in-house, constructive critics of the policies and operations of the AEC. Those same policies and operations became the province of the DOE and required the same constructive criticism.

The Joint House and Senate Committee on Atomic Energy had been a stern overseer of AEC decisions. Yet it also provided the necessary liaison between Congress and the public, given that the AEC would otherwise have functioned behind a veil of secrecy. Possibly most valuable of all, the joint committee stood between the AEC and the many congressional interests eager to grasp control of the agency.

The Military Liaison Committee brought to the attention of the all-civilian AEC still another point of view. Military personnel would have to devise the U.S. strategy in which atomic weapons would play a substantial part. They would deliver them to the enemy should the time ever come. Their concern was for adequate production and product efficiency. They represented the pragmatic outlook of the century-old military tradition, still important to national security in 1977.

Finally, the General Advisory Committee (GAC) represented the view of those at the heart of the enterprise, the scientists and engineers who had devised the reactors and the bombs. The GAC served to monitor the technical progress of the enterprise the AEC inherited and to advise it on further U.S. and international nuclear developments. The GAC was the vital technical body in the crucial international negotiations on nuclear weapons test bans and the verification procedures required of any test ban agreement. It insisted that the AEC sponsor applied research in its own laboratories to foster the continuing improvement of atomic weapons and the safe development of power reactors. Moreover, it insisted, with equal emphasis, on basic research in the new sciences that emerged after WWII to ensure that the United States did not fall behind in the science-dominated new world.

After the first rush of enthusiasm and approval of the creation of the DOE, it became clear that no energy policy would satisfy the contending forces in Washington. On one side were the proponents of a network of privately owned and operated nuclear power reactors as the solution to the nation's energy needs. These reactors, they argued, would in time replace the existing oil-and-coal-powered plants, eliminating much U.S. dependence on imported oil, and would help the country to enjoy a healthier environment. Opponents argued that the cooling required by nuclear power reactors was equally abusive to the environment and that the threat of accidents held hostage all who lived in proximity to a nuclear power plant. They offered natural gas, wood, and solar power as alternatives, at least in part, to oil, coal, and nuclear power.

In March 1979, two years after its birth, the DOE faced an accident at the privately owned and operated Three Mile Island nuclear power plant near Harrisburg, Pennsylvania. This event both fascinated and frightened the U.S. public during the several weeks required to secure the plant. It did not matter that no one was physically injured or exposed to anything more than a very small amount of radiation in the accident. Some months later, the presidential commission on Three Mile Island concluded that the accident was the result of "people-related problems and not equipment problems" and that "except for human failures, the major accident at Three Mile Island would have been a minor incident."[2] Nevertheless, Three Mile Island represented the end of growth for the U.S. nuclear power industry. The unease and outright panic generated in the public by Three Mile Island focused itself on the DOE, since it was the government agency purportedly responsible.

FIGURE 6.1. The Three Mile Island (tmi) power plant ten miles south of Harrisburg, Pennsylvania, showing the two cylindrical containment buildings (*center*) and two of the cooling towers (*background*) of the plant. The history of tmi is described in the newspaper article reproduced in figure 6.2.

Source: T. R. Fehner and Jack M. Holl, *Department of Energy, 1977–1994: A Summary History* (Oak Ridge, Tenn.: Office of Scientific and Technical Information, 1995), p. 27.

President Carter and his secretary of energy issued conflicting statements about the future of nuclear power in the United States. Following Three Mile Island, Secretary Schlesinger restated that the United States had "no real alternative . . . than to make effective use of nuclear power."[3] But the administration's second national energy plan, sent to Congress little more than one month after Three Mile Island, declared that during the past quarter-century the federal government placed a "disproportionate emphasis" on the nuclear production of electricity. President Carter also said that "we cannot shut the door on nuclear power for the United States" but added that once other energy sources were developed, "we can minimize our reliance on nuclear power which is the energy source of last resort."[4] Given this ambivalence, what then would be the DOE policy for nuclear power?

At last, a reactor that fulfills dream

Pa. was to be an atomic-power haven. Three decades and a disaster later, TMI Unit 1 fits the mold.

By Susan Q. Stranahan
INQUIRER STAFF WRITER

The world's most famous nuclear cooling towers are the legacy of a 1960s-era dream, promoted heavily by Pennsylvania business and political leaders, to convert the state into an atomic-power haven.

Under that scenario, the Three Mile Island reactors, built on a 2½-mile-long island 10 miles southeast of Harrisburg, would join other new reactors to supply so much low-cost, dependable electricity that business and industry would flock to the state.

Things didn't go quite as planned.

Long before the fateful morning of March 28, 1979, when residents along the lower Susquehanna River were awakened by a freight train-like rumble and a thunderous hiss of steam from two cooling towers that dominated the sky, there were plenty of signs that the Three Mile Island plant wasn't going to live up to those high expectations.

The cost of building the Unit 1 reactor had soared from its original estimate in the late '60s of $110 million to a finished price of $410 million in 1974. Its twin, Unit 2, would end up costing $700 million by the time it was completed in 1978, five years behind schedule.

That was only the beginning, however.

Three Mile Island soon became a global synonym for nuclear fiasco — environmental and economic. TMI were initials just about everyone knew.

On that chilly March morning, the main valve supplying cooling water to the glowing reactor core in Unit 2 had accidentally been shut off. Unbeknownst to control-room operators, the uranium fuel rapidly was overheating.

Emergency pumps kicked in, and water surged around the core. Steam formed, pressure in the containment vessel soared, and a relief valve clicked open — all according to design. But instead of closing when it should, the valve remained open and vital cooling water surged out of the 56-foot-high containment vessel.

The most dreaded of all nuclear accidents — a meltdown — was in the making. And nobody had a clue how to stop it. As conditions deteriorated, thousands of Pennsylvanians fled their homes in panic.

At the time, the Unit 1 reactor, a short distance away, was shut down for refueling and maintenance. It would remain shut down for 6½ years, as debate swirled in Harrisburg and Washington whether the reactors could ever generate electricity safely. Ultimately, more than $50 million was spent to upgrade Unit 1 before its operating permit was restored by federal authorities in 1985.

Despite early statements by TMI's owner, Metropolitan Edison Co., a subsidiary of General Public Utilities, that Unit 2 would one day be returned to service, the reactor was never restarted. More than $1 billion would be spent on the cleanup, which included removing tons of damaged fuel and heavily contaminated hardware. Unit 2 will be dismantled when Unit 1 is taken out of service.

Even while utility customers in Pennsylvania and New Jersey reeled as the cleanup costs and purchase of replacement power showed up in their monthly electric bills, and GPU struggled to return to fiscal health, Three Mile Island fooled the experts again.

The 800-megawatt Unit 1 reactor turned into The Little Engine That Could, producing reliable — and relatively inexpensive — electric power. Last year, Unit 1 set the world's record for uninterrupted service — 616 days and 23 hours.

That was the sort of performance those nuclear power boosters predicted three long decades ago.

FIGURE 6.2. TMI twenty years later. This July 18, 1998, article in the *Philadelphia Inquirer* recounts the history of the accident at TMI.

Source: Courtesy of the *Philadelphia Inquirer*.

Not long afterward, dissatisfaction with the DOE began to take form. It was directed at DOE involvement in the disappointing record of the nuclear power industry in general but especially Three Mile Island. And there were other reasons. The Carter energy policy, which the DOE was to implement and present to the public, was a curious mixture of inconsistent ideas. On the one hand, it spoke in favor of reliance on undeveloped energy sources such as solar energy and an enormously expensive investment ($88 billion) in a decade-long effort to improve the production of synthetic fuels from coal and shale oil reserves. On the other, the public was exhorted to conserve power in every aspect of its daily life. The ambivalence of the administration's attitude toward the nuclear issue, despite years of investment by the federal government in nuclear power, closed off that option and presumably led to the resignation of Secretary Schlesinger in July 1979, after two years in office.

President Carter quickly selected Charles W. Duncan Jr. to be the second secretary of energy. Duncan had a background in chemical engineering and management and previously had been deputy secretary of defense. The function of the DOE, as he saw it, was to carry out an energy program that was strictly defined by the national objectives set forth by the president. The DOE, he commented, should not be in the energy business. And he emphasized that "market forces must be allowed to regulate the price and allocation of energy resources such as petroleum."[5] Duncan began a tradition more or less faithfully followed by successive secretaries of energy. To streamline management and better delineate responsibilities for accomplishing DOE objectives, he moved the department toward a more traditional organization that managed programs by technologies or fuels. He discarded most of Schlesinger's philosophy and organizational programs.

In the presidential campaign of 1980, Ronald Reagan, the Republican candidate, advocated abolishing the DOE completely. He declared, "The DOE with its multibillion dollar budget had not produced a quart of oil or a lump of coal or anything else in the line of energy."[6] Nevertheless, in the midst of that management turmoil, the DOE continued to be one of the major research agencies in the nation. It owned and contracted out operation of the weapons research laboratories. It was the funding agency for several of the highest energy particle accelerator laboratories in the world, as well as of a host of smaller multidisciplinary laboratories, some with high-intensity research nuclear reactors. The research in those laboratories was unclassified and proposed and carried out independently by university and government scientists with no connection to weapons research. The DOE, through its Office of Energy Research, was thus one of the leading federal funding agencies for university scientists in the United States, both in number and dollar sums. The directors of the Office of Energy Research—equivalent to assistant secretaries in most government departments—were distinguished scientists, and as in the AEC and ERDA they required Senate confirmation for their appointments. They labored long and hard to promote science and technology from which universities, industry, and the entire nation benefited.

The DOE was submerged, however, in a multitude of chores and responsibilities, besides its function as a science agency, chores and responsibilities foisted on it in the interest of more efficient government organization. The secretary of energy, sitting in the president's cabinet, saw those other chores and responsibilities as the main business of the DOE. Each secretary of energy lived in fear of another energy crisis. And each secretary was the arbiter of deep differences of opinion between the administration and Congress concerning the government's proper roles in subsidizing the search for and development of new energy sources. Still, although preoccupied with those issues, the DOE did not desert its obligation to sponsor basic research. It simply gave less significance to that obligation while it concentrated on current business. Despite that difficulty, the science function of the DOE prospered. But as time went on, it acquired a foxhole mentality; it became more bureaucratic and more cautious and tended to micromanage its facilities and the research it supported. This mentality was in part a product of the attitude toward the DOE as a whole shared by the public and Washington. When Reagan advocated the abolition of the department, it was hard not to develop such a mentality.

The National Science Foundation budget passed a billion dollars while its directors came and went after brief service.

During President Carter's single term (1976–1980), he maintained that basic research was both a responsibility of and a wise investment by the federal government. In that period, the NSF's annual budget increased by sizable amounts, but high inflation and stagnant economic growth caused large federal deficits and severely limited real budget gains. Following Carter, the Reagan administration determined to continue and intensify the buildup of

U.S. armed services. This was intended to strain the USSR economically and militarily, since the two superpowers were still in cold war competition. As a result, federal deficits and military expenditures led to further retrenchments in nondefense spending.

In the mid-1970s, growing mistrust of the White House stemming from the Vietnam War and the Watergate affair led to increased congressional examination of the science establishment. Congress was suspicious of the integrity of the science agencies in the executive branch of the government and questioned whether what they were doing was worth the money they were spending. In the case of the NSF, questions started with the titles of grants and descriptions of funded research. Senator William Proxmire, chairman of the subcommittee with jurisdiction over the NSF budget, was especially critical of several grants in anthropology, sociology, and social psychology. He questioned the value to either the public or the government of such projects as "Hitchhiking—A Viable Addition to the Multimodal Transportation System" and "Social Behavior of Alaskan Brown Bears." Once again, the efforts of the NSF in those areas of the social sciences were deemed to be misguided and even harmful. This examination began near the end of the term of NSF director Stever, who vigorously defended the foundation against these congressional misgivings and attempted to counteract the false picture given to the public by emphasis on a few frivolous grant titles. Stever felt, however, that Congress was asking a legitimate question: what was the public getting for its money? He believed that both Congress and the public were entitled to a satisfactory answer and set about to provide it with NSF's grant application data.

Stever's testimony occurred at the time that the House Subcommittee on Science, Research, and Technology opened six days of hearings on the NSF peer review system. Peer review had been criticized ever since the earliest days of the NSF. Congressman John B. Conlan argued that the system was "closed and unaccountable to the scientific community and the Congress" and that "the NSF program managers could get whatever answer they want out of the peer review system to justify their [private] decision to reject or fund a particular proposal."[7]

The subcommittee, chaired by Representative James W. Symington, heard testimony from Congressmen Conlan and Robert Bauman, as well as Stever and his new deputy, Richard C. Atkinson. Conlan accused the NSF program directors of arbitrarily discarding negative reviews and purposely misrepresenting reviewers' comments. He advised the subcommittee "to make the peer review system open and accountable."[8] Bauman berated both Congress and the NSF. He recommended stricter congressional supervision of grant procedures by the authorizing and appropriations committees, in line with an amendment to the NSF bill he had previously put before the House. His amendment required the NSF to submit to Congress every thirty days a list of proposed grant awards, along with their justifications. Either chamber could line-item veto any grant award. Fortunately, the amendment did not pass.

These criticisms, initially prompted by the apparently frivolous titles and descriptions of a few of the NSF's awards, were not themselves frivolous. John Conlan, a graduate of Northwestern University and Harvard Law School, had been a Fulbright scholar in Germany and had taught at the University of Maryland and Arizona State University. His constituency had voiced concern about an educational project in the social sciences called "Man: A Course of Study" (MACOS), initiated under RANN, that the NSF had funded. As a course for fifth graders, it had reached seventeen hundred elementary schools in forty-seven states by 1975, when it was subjected to the charge that it severely distorted basic family values. The MACOS project had initially received favorable review by outside experts. It centered on the social habits of Netsilik Eskimos, but some of these were considered distasteful and ill suited for dissemination to schoolchildren in the lower grades. This concern caused Conlan to look more deeply into NSF grant procedures. Robert Baumann was also stimulated by local concern about MACOS and the expenditure of funds on what appeared to be foolish research.

Director Stever rebutted the charge that program directors manipulated the peer review system to benefit their friends. He insisted that all reviews of grant applications were required, as an agency rule, to be included verbatim in the application records. And he argued that the behavior of the system could be checked directly by assembling data that, when analyzed statistically, would show evidence of bias if it were present or, conversely, if it was not. He agreed, however, that the NSF should spot-check individual cases in the future, which it had not done in the past, and turned to Atkinson to present a statistical analysis of recent NSF grant performance.

Richard Atkinson had been chairman of the psychology department at Stanford University. He published extensively on mathematical models of learning and memory and was well equipped to make a quantitative statistical analysis of NSF data. Atkinson argued from the data that applications submitted by scientists from the top twenty departments in a given field had the same distribution of reviewers, geographically and otherwise, as applications from other schools. Nor did the eminence of the reviewers' universities correlate with the eminence of the universities from which the applications came. He concluded that the data he presented had "confirmed his faith in the fairness of the NSF review process" but that it was necessary for the NSF to collect data over a longer period of time and to explain more fully the working of the peer review system to the public and Congress.[9] Other researchers and administrators from outside the NSF also testified. In the main their views were completely consistent with Atkinson's presentation.

The report of the House Subcommittee stated that the NSF's "peer review evaluation systems appear basically sound" and that the NSF should continue to use them.[10] The report also recommended that the NSF attempt to achieve as much openness in the system as possible, but it firmly declined congressional review of individual research awards.

The Symington subcommittee report did not propose methods to open the peer review system. And, at a time when the nation was moving toward greater public access to the operations of the government as a reaction to Watergate, the NSF's position on the confidentiality of reviews and anonymity of reviewers continued to be questioned by critics who remained unconvinced of the fairness of its procedures. This issue was addressed by the National Academy of Sciences in a study of NSF practices. The NSF provided complete access to its records. Two professors of sociology not affiliated with the NSF did the study. The results were published in two parts, the first in 1978 and the second in 1981. No evidence was found for the existence of an old boys' network, but evidence of the high correlation between review ratings and awards was clearly demonstrated. Moreover, neither the age, race, or gender of the applicant nor his or her previous research accomplishments were found to have a negative influence on either the rating of an application or the probability of it receiving a grant.

To attack the question of personal bias more directly, one study was directed at evaluating the feasibility and promise of anonymous or blind applications, that is, applications in which the name of the author is suppressed. (Some institutions such as symphony orchestras had recently adopted such procedures and begun to hold auditions where the candidate was screened from the reviewers. In this way, the quality of playing and musicianship was the sole basis for judgment.) In its second phase, the study requested that program directors send 150 previously reviewed NSF applications to new reviewers. Half of them were edited to conceal the author's identity; authors were identified in the other half. The results of both surveys indicated that anonymous applications offered no clear advantage to the applicant or the NSF. No bias for or against any group was detected within either the anonymous or the author-identified applications.

Perhaps surprising, though not to experienced reviewers, was the result that about 25 percent of the funding decisions would be reversed if the applications were evaluated by another qualified group of reviewers. That finding attested to the substantive differences of opinion possible concerning the intrinsic value of an application and the importance of the area of science to which the application was directed. Reviewers also differed in their views as to whether the proposed work would be carried out successfully; they looked for originality of both purpose and method in an application. These attributes could be found to some degree by one reviewer and to a lesser degree by another, the former recommending approval and the latter, rejection. One pragmatic way to minimize the effects of these differences was to solicit evaluation from more than just a few reviewers—say, five to ten—which was a major final recommendation of the study. This way, perhaps in the final assessment the most negative review would be canceled out by the most positive one, as is done in judging some athletic competitions, such as figure skating or diving.

Examination of the NSF award system substantiated none of the accusations of bias or subjectivity. The system modified itself, however, in accord with the suggestions for improvement stemming from its self-examination. By 1977 the foundation routinely began sending copies of the reviews of their applications, on a trial basis, to investigators in the biological, behavioral, and social sciences. By 1983 that practice was adopted agencywide. It allowed all applicants to understand the basis for the funding decisions and provided information that made possible modification and resubmission of the original applications. The NSF also established a three-stage procedure for reevaluation of rejected applications but pointed out that funding did not necessarily follow review approval even of the first submission. The decision also took into account other factors, such as availability of funds, the relevance and significance to the NSF program from which the funds would flow, and the need to strengthen research throughout the nation. To reduce the complexity and number of applications, the NSF required that none exceed fifteen pages, and in 1980 it would reward effective, creative researchers with two-year extensions of their three-year grants without additional paperwork. To understand the process better, the NSF introduced an external oversight and review procedure of individual grants every three years. All these improvements became part of the NSF's methods, and a new Office of Audit and Oversight maintained records of activity. This allowed the NSF to justify its practices to Congress and to the public in periodic accountability hearings.

The peer review system was the core of the NSF's grant award procedures. Examinations of the NSF in the decade 1975–1985 modified the process toward greater openness and reception to applicant responses, but the process itself remained fundamentally intact. Despite the criticism of peer review elitism in the NSF, it remained thirty-five years later the fairest, workable method for the selection of good scientific research. No other system has ever been seriously proposed.

In 1981 and 1982 the Reagan administration cut the NSF's budget, especially in the areas of the social sciences and science education, which the Reagan White House believed were more properly supported by the states and the private sector. Nevertheless, in 1983, the third year of Reagan's first term, the NSF's budget passed the one-billion-dollar mark.

With its billion-dollar budget, the NSF was expected to increase innovative technology and engineering in its programs. It was argued that doing so would advance U.S. competitiveness worldwide. The NSF had raised engineering to a separate directorate in 1979, in which applied science programs were included. Two years later, the applied science programs were distributed to other directorates, but engineering was given a place alongside science in the science and engineering education directorate, just when the budget cuts for the education directorate occurred in the Reagan administration. At the same time, the engineering directorate established an office of interdisciplinary research to take advantage of collaborations among several science and engineering disciplines. Soon after that, an advisory committee recommended the creation of engineering research centers, each composed of voluntary groupings of scientists and engineers active in different but related science areas. Those centers were intended to facilitate cross-fertilization and possibly extend to technology transfers between universities and industry. Awards to six centers were made in 1985, ranging from a center for microelectronic robotics systems at the University of California at Santa Barbara to a center for biotechnology process engineering at the Massachusetts Institute of Technology. Apart from the recognized success of the centers, they served also to deflect Congress from submitting bills to create a National Engineering Foundation.

Despite favorable financial and scientific developments, however, a puzzling and questionable trend emerged within the NSF. The length of the terms served by successive NSF directors was alarmingly short. The leadership of the foundation changed frequently during the Nixon and Ford administrations; no director served more than half of a full six-year term: Richard Atkinson, previously deputy director under Stever, served three years, while John B. Slaughter and Edward A. Knapp each served only two years. Stability was restored in 1984 when Reagan appointed Erich Bloch, an engineer and former corporate executive, the first director to come from industry, who served a full six-year term. His immediate predecessors resigned to accept academic positions or to return to their professions. But a more likely explanation of the trend was that the challenge and sense of accomplishment offered by the directorship of the NSF was overshadowed by the stress and aggravation coming from the White House and Congress during that fifteen-year period.

The first linkup in orbit of Soviet and U.S. spacecraft occurred in 1975, and spacecraft shuttling between orbit and Earth became a regular feature of NASA's program; then came the Challenger *disaster.*

The decade 1975–1985 began with an important success for space flight, namely the linkup of a USSR SOYUZ spacecraft, already in orbit, with a U.S. *Apollo* spacecraft. In the two days spent together, crewmembers moved between the spacecraft, and the first concrete example of USA-USSR cooperation in space went smoothly. This was a remarkable feat given the cold war. Several years of joint planning, cooperation, and concern for the safety of the astro- and cosmonauts had been required. The people and governments of the world's two superpowers were aware of the broader implications.

The USA-USSR collaboration heralded a new era for NASA. The eleven-and-a-half-year preoccupation with the *Apollo* mission was over, and a self-confident NASA launched a variety of new tasks emphasizing Earth-oriented applications and basic science missions, all with international collaboration.

The space science program, *Viking*, would send missions to the planets of

FIGURE 6.3. Viking orbiter montage of 102 photos of Mars in February 1980 shows the Valles Marineris bisecting the planet, a gorge that would stretch from coast to coast of North America; to its left, three large volcanoes poke up through the unusual cloud cover.

Source: R. E. Bilstein, *Orders of Magnitude: A History of the NACA and NASA, 1915–1990*, NASA SP-4406 (Washington, D.C.: NASA, 1990), p. 112.

the solar system, first to the inner planets and later to the more enigmatic outer planets. Mars was the first target. *Viking* deployed four spacecraft in the vicinity of Mars, two orbiters to photograph the surface and serve as communication relay stations and two landers to descend to the Martian surface to measure the atmosphere and climate and search for evidence of rudimentary life forms. The spacecraft went into orbit around the planet in 1976, and subsequently the two landers descended safely to the rock-strewn surface. At that time, the planet was quiescent, but volcanoes half again as high as any on Earth and canyons deeper and longer than Earth's indicated a period several billion years earlier when Mars was active volcanically. Water was located in the frozen polar ice caps, but there was no evidence of life.

Venus was probed in late 1978. Its heavy, thick, hot atmosphere exhibited a high sulfur content with lesser amounts of oxygen and water vapor. The surface appeared to have two major continents and a massive island without an ocean, and there were mountains taller than Earth's Mount Everest.

In 1979 a new spacecraft system, *Voyager*, was sent to Jupiter with two spacecraft. Using Jupiter's gravitational field as a kind of slingshot, the two *Voyager* craft then set off for Saturn, where they arrived about a year and a half later. The mission was extended to a fly-by of Uranus in 1986 and to a planned fly-by of Neptune in 1989, if sufficient control fuel remained.

Studies of the Sun continued steadily also. Solar spacecraft received data about the effect of solar radiation on the earth's magnetosphere and the Sun's extraordinary eleven-year cycles. Part of this research was done jointly with the Federal Republic of Germany. Congress mandated a program to study the Earth's upper atmosphere to learn about the effects of gases such as freon on the ozone layer; this occupied NASA during the latter

FIGURE 6.4. *Top: Landsat 4* spacecraft photograph of New York City area in 1983. Images from the satellite were combined at the Goddard Space Flight Center. The island of Manhattan is near the center at the confluence of the Hudson and East rivers.

Source: R. E. Bilstein, *Orders of Magnitude: A History of the NACA and NASA, 1915–1990*, NASA SP-4406 (Washington, D.C.: NASA, 1990), p. 97.

Bottom: In the cutaway illustration, the shuttle orbiter is shown with the European Space Agency (ESA) Spacelab as the prime payload. Scientific instruments were mounted on ESA-built pallets arranged in the rear of the shuttle's cargo bay.

Source: R. E. Bilstein, *Orders of Magnitude: A History of the NACA and NASA, 1915–1990*, NASA SP-4406 (Washington, D.C.: NASA, 1990), p. 109.

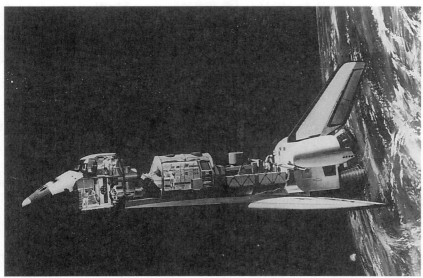

half of the 1970s. A profile and model of the ozone layer was the result. And *Landsat 3*, launched in 1978, continued the flow of worldwide data on Earth resources, collected mostly for the U.S. Department of Agriculture.

The space agency also remained true to its legacy of aircraft research. More efficient wing construction and improved fuel efficiency in jet engines influenced the construction of jetliners in the early 1980s. Other issues evaluated were aircraft noise during landing and takeoff, bad weather procedures, and control of high-density traffic patterns. The Ames Aeronautical Laboratory at Iowa State University began research on short-haul aircraft, especially vertical takeoff and landing (v/stol) aircraft. The laboratory also included flight testing and wind tunnel testing. Ames grew into nasa's leading center for helicopter research and contributed to research on tilt rotor aircraft.

In short, soon after the *Apollo* mission was completed, nasa had many irons in the fire. But the largest consumer of the nasa budget and management attention during the late 1970s was the space shuttle program. This $5.2 billion program included new designs of satellites and space flights that would carry academic scientists in addition to astronauts. The shuttles would carry payloads that could be placed in chosen orbits and retrieved. Shuttles could be reused many times. To make the cost manageable and the project salable, the shuttle would be launched vertically, jettison the solid-fuel booster rockets and the liquid hydrogen–liquid oxygen fuel tank, and return to Earth, landing like an airplane. The empty booster rocket casings that parachuted to Earth would be reused, but the fuel tank would burn up on reentry to the atmosphere. The shuttle was designed to carry a payload of sixty-five thousand pounds in orbit at 230 miles above the Earth and to accommodate up to seven crewmembers living and working in the flight deck area for long periods. Smaller payloads would allow orbits up to 690 miles. Nine years after the project had been approved by President Nixon, the shuttle *Columbia* went into orbit 130 miles above Earth for a two-day mission, the first of twenty-four missions by four different shuttles—*Discovery, Atlantis, Challenger*, and *Columbia*—in the following five years.

At liftoff, a shuttle looked and sounded like an oversized rocket booster with wings. It perched atop a cylindrical liquid propellant tank that fed the trio of main engines mounted in the shuttle's tail. A pair of maneuvering engines plus several small rocket thrusters refined the orbital path as needed during the mission. A shuttle in orbit was much larger than an *Apollo* spacecraft: it had a length of 120 feet and a wingspan of 80 feet. The cargo bay measured 60 feet in length by 15 feet in diameter. Shuttles were equipped with ceramic tiles over their outer surfaces to enable them to withstand the intense heat generated by air friction on reentry into the Earth's atmosphere. Some of these tiles worked loose during the flight of the first shuttle, *Columbia*, and occasioned some anxious moments among the mission controllers. At a speed of Mach 24, the shuttle entered the atmosphere and became enveloped by a blanket of ionized gases emitted by the white-hot tiles that disrupted radio communications. When *Columbia* slowed to Mach 10, it was cool enough to retransmit and reassure mission control that all was well. It was greeted at touchdown by an estimated half-million people who came to observe the "airplane" that had been in Earth orbit.

More than a thousand different payloads were proposed for shuttle space flights. Among them were several that were exclusively scientific, aimed at bringing into reality the observational and measurement capabilities that before had only been dreams.

In the early 1970s, NASA refurbished an airliner christened *Galileo* to carry out a variety of tasks such as observations in infrared astronomy, which at the time was a powerful new technique, photography of the Earth, and meteorological studies. The oceanic companion to *Landsat*, *Seasat*, despite a short life, provided information on the seas that had never before been available. But the shuttle space flights opened the way for complex scientific and communication equipment to be put in space, revisited for maintenance, and, if necessary, returned to Earth for modernization. The agency was on the threshold of new careers in space: a career in information gathering for civilian and military purposes; a career facilitating communication between far-removed points on Earth; and a career in science working in close partnership with academic research scientists. It planned a new family of observatories using shuttle-placed satellites that would carry equipment to study the broad spectrum of radiation from the most distant objects in the sky. A revolution in astronomy and astrophysics was in the making.

But before those space flights could be realized, a seemingly routine flight of the shuttle *Challenger* at the very beginning of 1986 became NASA's second tragedy. Seven Americans, among whom was a New Hampshire high school social studies teacher, Christa McAuliffe, were aboard *Challenger* when, about seventy-three seconds after liftoff, the shuttle exploded, destroying itself and the lives of its crew. It was a horrifying event that also destroyed the perception of take-for-granted success for every mission NASA conducted. Awareness was rekindled that travel into space was a dangerous business, requiring unceasing vigilance and attention to detail.

President Reagan appointed a special commission to conduct a formal enquiry into the tragedy. That report, released a few months later, was highly critical of NASA management and recommended that it be overhauled significantly. Technical evaluation led to the discovery of leaky booster joints that the commission held were the cause of the explosion. Richard Feynman, a Nobel Laureate in physics and member of the commission, showed by a simple experiment during one of the commission meetings that synthetic rubber o-ring seals forced to operate at very low temperatures were the cause of the leaky booster joints. The agency changed the booster design and introduced improvements in the shuttle's main engines, a crew escape system, and changes in other aspects of the shuttle's operations. It would be almost three years before the flight of shuttle *Discovery*, NASA's first manned mission after the loss of *Challenger*, would take place.

The *Challenger* disaster came twenty years after the loss of the lives of three astronauts, Virgil I. Grissom, Edward H. White II, and Roger B. Chaffee, in the flash fire that enveloped a spacecraft on the ground. Once again the issue of the price in human life and dollars that NASA was willing to pay to put astronauts in space was raised. It was argued that humans were not necessary for the scientific, economic, or military programs that were NASA goals because these could be accomplished with much less risk and much less cost by robotic instrumentation. It was claimed that NASA would have done better to invest in the development of robotics and eliminated the fragile human component of space flights. The money saved, said the critics, could have funded a number of programs with high scientific promise that would be either delayed or left undone because of the tragedy.

This issue is still hotly debated today with little prospect of resolution. The space agency functions differently from the other science agencies. It employs a large number of space scientists, astrophysicists, and astronomers. At the same time, it cultivates and supports university scientists, who have benefited enormously from the shuttle program but whose influence on NASA's long-range scientific policy is less than they would like. They find NASA's emphasis on costly nonscientific space flights—particularly on humans in space—to be superfluous and wasteful.

But NASA is in the public eye in everything it does. Ever since the *Apollo* program put men on the moon, NASA has been associated in the public mind with the romantic idea of humans risking themselves in the challenge to explore space, a new but marvelously mysterious and attractive frontier. Even scientists, when asked what the twentieth century will be remembered for, recognize that high on the list—possibly first—will be the human escape from the Earth to visit the moon, the first of many escapes to space that are likely to occur in the next millennium. But NASA has not yet found an effective way to resolve the principal issue raised by scientists: that it does not listen to them as well as it might.

The National Institutes of Health no longer had sufficient funds for the rapidly increasing number of research grant applications; directors were dismissed, and morale plummeted.

The science agencies of the federal government breathed a collective sigh of relief when Nixon resigned his administration, and none was more heart-felt than that of the NIH. The institutes were in the midst of another revo-lution in biomedical research, particularly in cell biology and recombinant DNA, and in their many startling clinical applications. These stimulated a corresponding increase of competing grant applications by almost a factor of two between 1973 and 1978. This deluge of applications came at a time when budget cutbacks in the overall medical sector were motivated by the perceived danger in the runaway Medicare/Medicaid budgets. The combi-nation of increased demand for grants and decreased resources as a result of high inflation perpetuated the problems that the NIH had experienced in the previous decade. Both Congress and the grant applicant community believed, each for its own reasons, that the fault lay with the NIH manage-ment. Once again, Congress set out to provide better management proce-dures by legislative directives. The applicants, feeling desperate about what they thought to be an outrageously low approval rate, sallied forth, often with vehemently expressed suggestions for changing the system, even including the elimination of peer review.

The one issue on which Congress and applicants were agreed was the dis-tressing complexity of the application procedure. Congress reacted because of complaints from constituents; applicants complained because of personal experience. There was some justification for their criticism. Soon after it was created, the Division of Research Grants established a peer review system intended to provide study section reviewers with all the information they might require to make a quick, informed, and balanced recommendation for each and every grant request. Over time, the applications became significantly more complicated; for example, some required additional information to satisfy federal and state laws governing the use of radioactive materials, the handling of laboratory animals, and myriad other details about the anticipated support to be furnished by the host institution. The number of pages and number of copies of an application needed for the many member review sections had become an expensive, time-consuming burden for both the applicants and the DRG. The DRG tried to lighten the load on applicants by organizing tutorial sessions on application procedures. But it was caught on the horns of a major dilemma. On the one hand, for the purposes of accountability within the NIH and to Congress, elaborate records of the reviews of each application had to be maintained for long periods. Statistical analysis of the numbers of applicants, their institutions, their geographical locations, and so forth, were likewise necessary for the DRG's yearly presentation to Congress. Moreover, the analysis had to be up-to-date to satisfy inquiry at any time by a member of Congress. On the other hand, the DRG was unwilling to prejudice the peer review system by shortcutting any but the most innocuous of its requirements. The result in the restrictive economic climate of the Carter and early Reagan administrations was a stalemate.

The political autonomy of the NIH was compromised when outsiders took sides. Attempts to defend the integrity of the biomedical research establishment led to the Nixon administration's summary firing of the NIH's director, Robert Q. Marston, in 1973 and forced the resignation of another NIH director, Robert S. Stone, in 1975. A decade later, the situation of the NIH was not much better; it was still unstable and unpromising.

In 1982 William F. Raub, the NIH associate director for research and training, had written a strategy paper containing this passage:

> During the last few years, there has been a slowly spreading realization within the biomedical research community that the enterprise not only has stopped growing but actually has begun a contraction of unpredictable duration. Competition for funds from NIH and other sponsors, intensifying year by year, now stands at an unprecedented level, and shows no signs of abating. Never before have so many established investigators faced so much uncertainty about their longevity as active scientists. Never before have so many novices faced so many disincentives to entering or continuing a research career.[11]

FIGURE 6.5. *Top*: Dr. James B. Wyngaarden, director of the NIH, 1982–1989.

Source: Richard Mandel, *A Half Century of Peer Review (1946–1996)* (Alexandria, Va.: Division of Research Grants, National Institutes of Health, Logistic Applications, 1996), p. 180.

Bottom: Dr. Antonia C. Novello, executive secretary, General Medicine B Study Section, 1981–1986; surgeon general, U.S. Public Health Service, 1990–1993.

Source: Richard Mandel, *A Half Century of Peer Review (1946–1996)* (Alexandria, Va.: Division of Research Grants, National Institutes of Health, Logistic Applications, 1996), p. 188.

Five years later, in a briefing on peer review, the NIH's director, James B. Wyngaarden, described the process as follows: "It is no myth that the pressure for greater accountability for the use of federal funds has (1) made the grants application process more burdensome for investigators, university administrators, and members of peer review groups; (2) contributed to additional uncertainty and insecurity in the careers of extramural scientists; (3) created impediments to the creativity and productivity of investigators."[12]

These assessments described the dreary and foreboding attitude of the NIH as it made ready for the last decade of the twentieth century. Nevertheless, on the bright side, there was a commitment to at least five thousand new and renewed investigator-initiated awards at the then-current level of federal funding. Study sections, despite being overworked and understaffed, were reviewing successfully large numbers of applications per year. And from 1980 to 1989 funding for extramural awards increased from $2.8 billion to $3.5 billion, adjusted for inflation. Once again, the NIH overcame its continuing troubles and survived, the peer review system along with it. Moreover, the NIH emerged as the preeminent world institute for the health sciences.

The close of the decade sees tighter management and overall expansion.

The nation that the Carter and Reagan administrations inherited was politically turbulent and economically distraught. A natural reaction of each administration and Congress in those periods was to look critically at various parts of the federal government and, where possible, to modify the principles and practices under which each was operating, much as an individual who has lived through a traumatic experience turns with relief to the job of restoring order and efficiency in his or her own life.

The urge to put its house in order took a different form in each of the federal science agencies. For the AEC, it meant absorption into the Department of Energy. For the NSF, there were increased demands to direct its programs toward greater relevance to national needs. The NSF also undertook self-scrutiny of its award practices to appease congressional demands. Following the history-making achievements of the *Apollo* missions, NASA consolidated its many interests and focused on the space shuttle program, which brought it closer to collaboration with academic scientists. The NIH was caught up in government concern over the rapidly growing cost of medical care and plagued by increasing fiscal and procedural constraints. Within the ranks of the U.S. medical research community, there were strident cries to an overburdened, overexpanded system for more support for more investigators.

Despite this woeful litany, the science establishment managed to expand in the decade 1975–1985. The establishment proved to have the stamina and flexibility to survive the stresses in its evolving relationship with the government.

CHAPTER 7

Golden Anniversary: 1985–1995

As the golden anniversary of the marriage approached, the compact between the science establishment and the federal government remained intact and as felicitous as long-term compacts between the government and its citizens are likely to be. During those fifty years, the U.S. experience also helped to define science support in industrially developed nations everywhere. The essential feature adopted by the United States and many other nations was support of peer-reviewed proposals for basic research by individual scientists.

By 1995 the U.S. science agencies had matured in their role as intermediaries between the government—the Congress and the president—and scientists. Important similarities and differences in character among the individual agencies showed up more clearly than before. Just as individual men and women age differently, some more fortunate in their lives than others, some welcoming and flourishing under change and some resentful and unaccommodating, so did the individual science agencies. The science function of the DOE was submerged under other tasks, and change was not welcome. In contrast, NASA continued the transition from its manned space flights to a period emphasizing broadly based science and technology.

While continuing to serve as the universally acknowledged Mecca of biomedical research, the NIH was caught up in the general health care dilemma facing the nation. Finally, the NSF, like the NIH, emerged from a period of turmoil in which it successfully refuted accusations of incompetence and favoritism. The NSF consolidated its position as principal supporter of basic science and, without fanfare, extended support to developing areas of new science.

The Department of Energy endured undeserved backlash from Chernobyl, and problems of nuclear waste management and environmental cleanup of its own laboratories exacerbated its woes. Then with physicists' help, it bungled the Superconducting Super Collider (particle accelerator) project.

In October 1997 the DOE celebrated its twentieth birthday. During that twenty-year period, the United States had four presidents and eight secretaries of energy. The problems and events that the DOE faced involved mixtures of complex technical and political issues, some with straightforward solutions, some not.

In 1986 the nuclear power plant at Chernobyl in the Soviet Ukraine overheated and exploded, dispersing large amounts of radioactive material over all of eastern Europe and Scandinavia, and as far west as the nations on the Atlantic coast of Europe. The DOE reacted immediately to the meltdown, which was enormously more serious than the accident at Three Mile Island. It arranged to send reactor specialists from its own laboratories to help with containment of the still smoldering wreck and to safeguard the remaining reactors at the site. And it called on specialists in nuclear medicine to help with the treatment of severely irradiated plant workers and others affected who lived nearby.

The DOE had nothing to do with the miscalculations that produced the disaster at Chernobyl. The reactor was designed differently from U.S. reactors. Its design—including grossly inadequate safety controls and insufficient building containment of radioactive material in the case of a possible reactor accident—would never have left the drawing board in the United States, much less have been built. Furthermore, DOE scientists and engineers had demonstrated well before Chernobyl that nuclear reactor safety could be ensured by proper reactor design. Its Civilian Reactor Research and Development Program worked on the development of passively safe

nuclear power plants and demonstrated that certain types of reactors, operating at full capacity, would automatically shut down when all cooling systems ceased to operate. The explanation for this automatic shutdown was
that the natural laws of physics, not engineered safety systems, kept reactor
core temperatures within safe limits and provided passive, as opposed to
active, safety. In the cold war climate of fear of the time, this feature was not
widely advertised. Nor were the critical differences between Russian power
plants and U.S. power plant reactors made clear to the American public. As
a consequence, little credit went to the DOE or to the expertise of its scientists and engineers for the accomplishments of its laboratories in the development of safe nuclear power.

A different aspect of nuclear power that also plagued the DOE was management of high-level nuclear waste from its own and privately owned
nuclear reactors. The Nuclear Waste Policy Act of 1982 enjoined the DOE to
site, design, construct, and operate the first U.S. geologic repository for permanent disposal of spent fuel and high-level waste from civilian nuclear
reactors. Four years later, President Reagan selected three sites, all in western states, for study by the DOE; one of them would be recommended as a
permanent site. Congress short-circuited this procedure with the Waste
Policy Amendments Act of 1987 that designated the Yucca Mountain site in
Nevada as the only candidate site to be considered. The governor of Nevada,
Richard Bryan, and Nevadans in general strongly opposed that decision, on
the grounds that Nevada had been the site for years of the federal government's underground nuclear weapons test facility and needed no further
radioactive waste within its boundaries. Two years later, the next governor
of Nevada, Robert Miller, also outraged by the decision to concentrate
solely on the Yucca Mountain site, signed into law a bill that made the storage of high-level radioactive waste in Nevada illegal. So began a contest
between the DOE and the state of Nevada in 1989, a case that went to the U.S.
Supreme Court. The Supreme Court decision, as observed by a spokesman
for Richard Bryan (by then a senator from Nevada) was "just one skirmish
in what has been and will be a long battle."[1]

Two other waste storage battles occupied the DOE during the same
period: the Monitored Retrievable Storage (MRS) site and the Waste Isolation Pilot Plant (WIPP). The MRS was mentioned in the Nuclear Waste Policy Act as an interim storage site in which regular monitoring was possible;
when a permanent site became available, the waste material would be
moved. The act required identification of a state or Native American tribe

amenable to hosting an MRS facility. As of February 1992, the DOE had received seven applications for grants to study the prospects of an MRS location; however, no action of significance followed.

As for the WIPP, the DOE spent an average of $100 million for each of the seven years it took to construct a facility near Carlsbad, New Mexico, the region of the famous Carlsbad Caverns. Again, before radioactive material—mostly, spent fuel cells—could be deposited there, the issue landed in the courts. In this case, however, the DOE prevailed after twenty-five years of intensive on-site studies, protests, and lawsuits. In the spring of 1999, the $2 billion Waste Isolation Pilot Plant began receiving material for storage.

The DOE faced another serious technical problem. It needed to clean up the long-lived radioactive material scattered throughout the laboratories and isotope separation plants that had been the centers of uranium and plutonium production during WWII. During the war the standards for radiation safety were much looser. Once the harmful effects were better understood, radiation exposure limits were made far more stringent. Advances in nuclear medicine and case studies of bomb victims showed how the human body reacted to specific radioactive elements, such as the sensitivity of the thyroid gland to radioactive iodine and of the lungs to radioactive strontium. Those studies also led to stricter standards for external human body exposure to radioactivity. In short, radioactive cleanup was a technical problem that the DOE was nominally well equipped to handle.

In 1985 responsibility for DOE environmental, safety, and health programs was consolidated under a newly created assistant secretary. A year later, a special committee of the National Research Council conducted a survey of technical environmental safety at more than fifty DOE facilities. Among other findings, the committee discovered a surprising situation: the DOE lacked adequate technical understanding and capability to handle the problem. Equally serious was the conclusion that "weaknesses of management had led to a loose-knit system of largely self-regulated contractors."[2] John S. Herrington, secretary of energy under Ronald Reagan, promised action and established an independent oversight panel to propose corrective plans. The panel's study, reported in July 1989, after Herrington had left office, focused on seventeen sites and estimated expected cleanup and environmental compliance costs to be $66 billion through the year 2025; but a high estimate went to $110 billion by 2045. Senator John Glenn, former astronaut and chairman of the Governmental Affairs Committee, characterized the high estimate as likely to be the floor, not the ceiling.

The Bush administration named Admiral James D. Watkins to be secretary of energy in 1989. He left office three years later. He provided a retrospective of his tenure at the DOE, stating that his foremost accomplishment was implementation of "a new management culture that understands the need for compatibility between our defense mission and protection of the environment."[3] According to Watkins, the DOE had given first priority to bringing all facilities into environmental compliance. He admitted that at the end of the cold war in 1990, the DOE was not capable of producing new nuclear weapons. If it had been required to do so, it would have had to ask President Bush to override safety and environmental laws to resume production at facilities that would have been "safe enough, but not at a desirable level."[4]

During the first Clinton administration, the DOE, then under Secretary Hazel O'Leary, was spending $6 billion annually, fully one-third of its budget, on a still coalescing program of facilities cleanup. That program was described by Senator J. Bennett Johnston, chair of the Senate Energy Committee, as a "grand and glorious mess."[5] The new DOE assistant secretary for environmental management, Thomas P. Grumbly, acknowledged the lack of any concrete results. He explained "that everything we do is driven by compliance agreements."[6] These were essentially cleanup blueprints specifying enforceable milestones at each site, but at many sites the problems were "larger, more complex or simply different than we had originally expected."[7]

All this tested the DOE management. The results were mixed. Where nuclear reactor safety was concerned, the AEC and DOE had good records: no government-operated nuclear reactor had been a source of any trouble whatsoever. The many privately operated U.S. power reactors with reactor designs consistent with AEC specifications had good records except for the accident at Three Mile Island, which was caused by human error and, more important, did not result in any physical harm. On the other hand, the DOE record in waste management and environmental cleanup at its own facilities was dreadful.

Yet the solution of radioactive waste management—how and where to store spent fuel rods from the DOE's own and private power reactors—was essentially a technical one, and completely feasible. The DOE soon settled on several methods of packaging the radioactive material, for example, sealing it in thick-walled, initially molten glass cylinders encased in metal containers. The cylinders were subjected to stringent mechanical shock tests and seepage tests that would indicate even very slow oozing of the encased material through the glass and metal. The conclusion was that no measurable amount of radioactive material would leak from the container in less than ten thousand years, probably longer. Assuming that the cylinders would be stored in deep underground sites well engineered and carefully selected for geologic stability and deep water tables, the DOE technical staff believed that it had produced a technical solution for the waste management problem. A facility to test the method was planned under Yucca Mountain in the Nevada Weapons Test Site. The last of the necessary permits to build were obtained in March 1992.

Once again, however, the reaction of Nevadans and citizens of other western states was extremely negative. And the response of Congress to the issue was mixed and weak. The DOE did not exhibit the conviction of purpose or the continuity of leadership necessary to deal effectively with the stalemate. None of the eight secretaries of energy, whose average tenure was less than three years, was able to convince the public and Congress that, whatever the long-term future of nuclear power in the United States, the problem of spent fuel cell storage was serious—many reactors were in operation, and the problem would only become more serious if neglected—and that the DOE had sensible, tested solutions to the waste management problem that needed only to be implemented. Furthermore, most new secretaries tended to denigrate publicly the internal organization and lack of accomplishment of previous secretaries, and this reinforced congressional and public skepticism of any waste management plan offered by the DOE.

The cleanup of DOE facilities did not involve public or congressional approval, apart from its large cost in the department budget. It was strictly a technical problem, but a successful solution was not forthcoming.

Perhaps the most telling failure of the DOE, however, aided and abetted by university scientists, involved a project to construct by far the largest, most energetic, and most expensive particle accelerator in the history of physics: the modestly named Superconducting Super Collider (SSC), designed to produce accelerated particles, subatomic in size but with energies rivaling any found in the particle spectrum of natural cosmic rays. The highest-energy cosmic rays, composed mostly of protons, are thought to originate in deep space and to be accelerated by the weak magnetic fields in space during the eons of travel time it takes to reach Earth. The SSC was intended to accelerate protons to energies comparable to those in cosmic rays by using a combination of very intense magnetic and electric fields.

The data acquired from experiments at lower-energy accelerators and the theories developed to explain the data strongly suggested that new phenomena would be present at ssc energies. It was claimed that experiments at ssc energies would revolutionize our understanding of the elementary particle world and perhaps also cast light on the origin of the universe.

Physicists had made similar claims ever since they had convinced the AEC—soon after WWII—to fund high energy physics (HEP) and the particle accelerators required for its study. With time, the need for accelerators of higher and higher energy led to the commitment by the AEC and its successors of larger sums for the construction and operation of new accelerator facilities. For example, the AEC paid for the highest-energy particle accelerator in the world and its associated facilities at a cost of $240 million in the early 1970s. That accelerator complex, located outside of Batavia, Illinois, and named Fermilab, was also funded annually by the AEC. It was managed, however, by a consortium of universities with interest in HEP. While Fermilab had a large staff of physicists and engineers, the accelerator was used primarily by university physicists whose proposals for experiments needed favorable peer review before they could be constructed and put in place. In addition, the university physicists' work was funded annually through AEC contracts—specified for HEP research—with their respective universities.

The reasons given to justify the large amount of funding were various. Some in Congress thought that national defense was the primary reason, some that U.S. stature in the international world of science was justification enough, and some that research in this far-out field was an investment in the unknown—recommended by many of the nation's most accomplished scientists—that the United States could ill afford not to make. Scientists in other fields had a variety of opposing opinions concerning the contributions of HEP to the national defense and the quality of American life. But negative views of high energy physics and its cost did not deter the university professors and students who were engaging in research in the subject. They saw themselves as seeking the "substance of substance," the elementary particles of which all matter is made: in a word, probing nature at its most fundamental level.

When the cold war was at its height, a congressional committee asked Robert R. Wilson, a former researcher at Los Alamos in WWII, professor at Cornell University, and director of Fermilab, to explain how high energy physics aided the defense of the United States. He turned the question on its head by responding that the freedom to pursue research in high energy

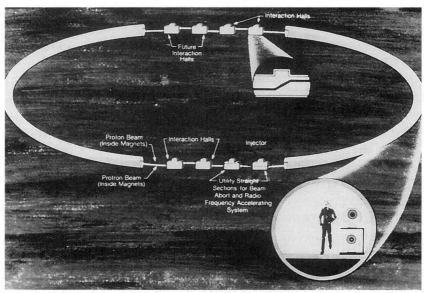

Future
Interaction
Halls

Interaction Halls

Proton Beam
(Inside Magnets)

Interaction Halls

Injector

Protron Beam
(Inside Magnets)

Utility Straight
Sections for Beam
Abort and Radio-
Frequency Accelerating
System

physics—the freedom to study at the farthest reach of the frontier of science—was one of the many freedoms that made the United States worth defending. It was that sense of mission that motivated high energy physicists and led them to propose the ssc.

The DOE encouraged the high energy physics community. In 1983 the High Energy Physics Advisory Panel (HEPAP) OF THE DOE and the director of the Office of Energy Research, Alvin W. Trivelpiece, recommended that it be assigned the highest priority. The project was expected to strain the federal research budget, but President Reagan's science adviser, George A. Keyworth II, and the secretary of energy, John S. Herrington, subsequently endorsed it. There was no disputing the potential scientific value of the ssc, but serious questions were raised concerning its impact on the funding of other areas of U.S. science. These divided the university science community.

President Reagan approved construction of the ssc in January 1987, when Secretary Herrington observed that it was equivalent "to putting a man on the Moon,"[8] a statement that did not endear the project to many scientists who were more skeptical of its importance. The total project cost was estimated at $4.4 billion over about ten years, based on a design study carried out by a multiuniversity team of accelerator experts in residence for several years at the University of California at Berkeley. The DOE proceeded to develop a site selection procedure, which, for such an expensive federal project, was typically a delicate business, usually making more enemies than friends. The ssc was no different in this respect. When the location choice finally settled on a site in Texas near Dallas, the enthusiasm of a number of political proponents of the project from other states waned. Nevertheless the project marched forward: the Universities Research Association (URA) was selected to manage construction of the accelerator and

FIGURE 7.1. *Opposite page top*: Department of Energy Secretary John S. Herrington (1985–1989) and Texas governor William Clements viewing an artist's conception of the Superconducting Super Collider (ssc) in 1988.

Source: T. R. Fehner and Jack M. Holl, *Department of Energy, 1977–1994: A Summary History* (Oak Ridge, Tenn.: Office of Scientific and Technical Information, 1995), p. 45.

Opposite page bottom: Schematic outline of the ssc showing the fifty-three-mile circumference tunnel and location of the related facilities.

Source: T. R. Fehner and Jack M. Holl, *Department of Energy, 1977–1994: A Summary History* (Oak Ridge, Tenn.: Office of Scientific and Technical Information, 1995), p. 45.

the new laboratory (the URA had successfully managed the construction and operation of a similar but smaller accelerator at Fermilab in Illinois). A project director was chosen, and the original design group was ready to go to work at the Texas site.

Progress on the conventional construction was made in the first few years. Strangely, the original design group was disbanded without explanation. By mid-1987, however, the cost estimate for the SSC had risen to $5.9 billion. Then, in January 1991, the DOE informed Congress that the new cost estimate was $8.25 billion, almost double the amount approved by President Reagan in 1987. Secretary O'Leary pledged in August 1993 that the cost would be held to $8.25 billion plus $2 billion in "stretch-out funding" to account for delays. One month later, the ante had risen to $9.94 billion plus stretch-out costs, according to a seventy-five-member committee headed by the DOE's procurement officer. In October 1993, in a last-minute effort to avert termination of the project by Congress, Secretary O'Leary informed the House that the cost of the project was tentatively estimated at less than $11 billion and would be held to that limit or new options for its fate would be presented by the DOE. Later in the fall of 1993 Congress terminated funding for the SSC project, leaving behind roughly five years of construction— a fifty-two-mile-circumference tunnel, laboratory buildings, and Texan farm and home land that had been bought by the state for the project. The dreams of the world high energy physics community focused on the SSC as the premier scientific instrument of the era were swiftly and thoroughly dashed. On the other hand, there was little mourning among scientists in other fields. They saw this outcome as a fitting response to the prideful attitude of the high energy physicists.

This had never happened during the fifty-year partnership of the physics community with the AEC, ERDA, and the DOE. A much smaller accelerator project had been terminated by the DOE almost two decades earlier, but the reason given for cancellation was inadequate progress on the superconducting magnets needed for the accelerator design. The demise of the highly advertised, high-priced SSC was a much different matter. It reduced the stature of the entire science community and the DOE throughout the government and widened the fissure between scientists and the DOE. An adequate account of the SSC has not yet been written, but it is sure to be a story full of ambition, intrigue, and human flaws.

A cursory explanation assigns culpability more or less equally to the high energy physicists involved and the DOE. Internal dissension among the physicists led to the dispersal of the original design group, an early sign of personality clashes and disputes at the director's level. Congress had been assured that 20 percent of the initial cost of the ssc would be contributed by foreign governments, but the necessary diplomatic and negotiating skills to acquire those funds were lacking, and the ever-increasing cost led to diminished enthusiasm abroad for the ssc. In the end, less than 3 percent of the original cost came from abroad.

As the estimated cost grew out of hand, both the physicists and the DOE exhibited rigidity in their behavior that was ill suited to a project of the magnitude of the ssc. The physicists did not propose feasible modifications of the original scope of the project, modifications that would still have yielded a valuable scientific instrument but at a much lower cost. They stubbornly insisted on all—no matter what the cost—or nothing. The DOE attempted to remedy the situation by taking control of the ssc project from the physicists and the URA management in all respects but name. A DOE contingent of several hundred people from its Washington headquarters was relocated at the ssc site to monitor commitments, expenditures, and construction progress. Soon, the DOE and the scientists grew contentious. Termination was inevitable.

On a completely different note, however, the DOE played an important, successful part in the origin of federal support for the human genome project. The DOE had for many years sponsored research in several of the laboratories it funded on the biological effects of radiation, especially genetic mutations. In 1983 the Life Sciences Division at the Los Alamos National Laboratory established a major data storage facility for genetic information. Known as Genbank, the facility obtained and stored DNA sequence data.

The director of the DOE Office of Health and Environment in Germantown, Maryland, was Charles DeLisi, formerly chief of mathematical biology at the NIH. Interested in how the data in Genbank might be used to study the genetic bases of human diseases, DeLisi speculated on the feasibility of acquiring a data bank containing the base-pair sequences of an entire human genome. At about that time, Robert Sinsheimer, a distinguished molecular biologist and chancellor of the Santa Cruz campus of the University of California, had the same idea. Independently, they organized workshops in Santa Cruz, in 1985, and Santa Fe, New Mexico, in 1986. Most of the participants were leaders in developing the methods and studies required to carry out the huge task presented by the human genome; in their view, the technical capability to do so was available.

Soon thereafter, enthusiasm for the human genome project and federal support of it began to mount, despite its size—enormous for biology—and cost—large for any scientific discipline. Charles DeLisi was in the forefront of the enthusiasts. Moreover, he spoke for the DOE and the $4.5 million allocated for the project in the DOE's fiscal 1987 appropriation. He advanced a plan for a five-year DOE program that made use of the technical strengths of the DOE laboratories. Toward the end of 1987 the secretary of energy ordered the establishment of human genome research centers at three of the DOE national laboratories: Los Alamos, Lawrence Livermore, and Lawrence Berkeley.

A number of biological and medical scientists questioned the fitness of the DOE—traditionally dominated by physical scientists—as the control center of the project. They were also distressed by the absence in the plan of the NIH, the principal federal agency concerned with the life sciences. These and other considerations brought about the support of James Wyngaarden, the director of the NIH, for a substantial NIH role in the human genome project. In December 1987 Congress appropriated approximately $17 million to the NIH and about $11 million to the DOE for human genome research in fiscal 1988.

So began the U.S. federally supported human genome project, appropriately shared between the two federal agencies with interests and talents vital to its success. The part played by the DOE, through DeLisi, in advancing the project in its early stage was salutary, forward-looking, and responsive to the needs of both the DOE and the science.

Nevertheless, the DOE as a whole has not been a successful agency. No secretary of energy has been able to organize it, to provide internal stability or solutions to the problems it has faced. Between 1977 and 1995 it sustained five major internal reorganizations at the hands of new secretaries of energy, each shifting or reversing the effect of an earlier reorganization. The DOE management of radioactive waste disposal and cleanup of its facilities has been at best inadequate. And DOE mismanagement of the ambitious scientific SSC project marked a low point in the history of the science establishment.

Although the magnitude of the research funds for which it is responsible is large—almost three times larger than the NSF—the Office of Energy Research of the DOE does not have the intellectual standing within the scientific community that the NSF has or, for that matter, that NASA has acquired in recent years. The DOE has been a study in inconsistency on the part of one administration after another, one Congress after another, and one energy secretary after another. All have contributed to making the DOE a catchall of energy issues and problems in technology without any ready solutions. Of the four major, civilian federal science agencies, the DOE has become the one most in need of substantial repair.

NASA recovered from the Challenger *disaster and concentrated on deploying many Earth-orbiting satellites.*

It took three years after the 1986 *Challenger* disaster before NASA and the space shuttle program recovered. That period of introspection and self-criticism affected all subsequent launches and space flights and brought about redesign of many shuttle components. In the interim, a number of unmanned space flights propelled by other launch vehicles were attempted. But several of these failed and added to NASA's sense of discouragement. In May 1986 a Delta rocket carrying a weather satellite was destroyed in flight after a steering failure. A year later, an Atlas-Centaur rocket for the navy's launch of a fleet satellite communications spacecraft was struck by lightning and broke up less than a minute after liftoff. A few months later, three rockets at the Wallops Island facility were ready when the launch pad was struck by lightning and all three shot off and crashed into the sea. And one month after that, yet another Atlas-Centaur rocket was destroyed by an industrial accident on its Cape Canaveral launch pad.

The space administration badly needed a centerpiece program for its own and public morale. In September 1988 the first post-*Challenger* shuttle flight took place successfully, and the shuttle program resumed without incident. In the seven years following, NASA placed forty-four satellites for industrial communications, thirteen weather observation satellites, and twenty-seven satellites devoted to the global positioning system (GPS) into Earth orbits. The communication satellites were an integral part of the revolution in information transferal that took place in the early 1990s, making the World Wide Web possible. The weather satellites extended and refined the weather database and its predictive precision, and the GPS satellites provided the coordinates of a point anywhere on Earth.

Another twenty-seven satellites were devoted exclusively to advanced scientific enquiry, of which eleven required the large cargo bay capacity of the space shuttle. That excursion into basic studies in astronomy, astrophysics, and cosmology brought NASA into far deeper and more extensive collaborations with university scientists than before. At the same time, the

FIGURE 7.2. Launch of the space shuttle *Discovery* and its five-man crew on a four-day mission to deploy the Tracking and Data Relay Satellite. Crew members were Commander Rick Hauck, Pilot Richard Covey, and mission specialists Dave Hilmers, Mike Lounge, and George (Pinky) Nelson.

Source: Richard Mandel, *A Half Century of Peer Review (1946–1996)* (Alexandria, Va.: Division of Research Grants, National Institutes of Health, Logistic Applications, 1996), p. 146.

sophisticated science satellites launched by the shuttle opened new windows to the universe through which university scientists had never looked.

The most widely known of the science satellites is the Hubble space telescope (HST), launched in April 1990 and named for the astronomer Edwin Hubble, who first suggested in 1928 that the universe is expanding. The HST is best known for two reasons: the breathtakingly beautiful high-resolution color pictures of stellar bodies that it routinely obtains and sends back to Earth for display on the Web and the mistake in the initial preparation of the ninety-four-and-a-half-inch-diameter mirror, the heart of its optical system. Critics complained about the expensive mistake, but the public loved the drama of the 1993 shuttle flight repair job. The blurred images recorded before the mirror aberration was fixed helped to suggest a corrective procedure and to indicate the replacement components. The shuttle *Endeavour* delivered the components to the orbiting HST, and two *Endeavour* astronauts who had practiced on a model back on Earth refitted the mirror with almost no trouble. The success of the repair was quickly verified when scientists looked at a few images sent back to Earth before *Endeavour* returned home. This experience and others in which satellites, such as the Russian space station *Mir*, were repaired while in orbit foretold one aspect of NASA's future: a space station furnished by means of shuttle payloads. Ultimately, the space station would be the transfer point for establishment of a human colony for scientific and technological studies on the Moon.

The advantage of the HST over ground-based optical telescopes is that its spatial resolution is not limited by the Earth's thick, constantly changing atmosphere. Astronomers have dreamed of this since the beginning of modern astronomy. The HST sees objects seven times further away and more clearly than any Earth-bound optical telescope. It's so-called deep field survey has allowed astronomers to study the structure of galaxies that are closer to the edge of the visible universe, and in doing so the HST has filled fundamental gaps in our knowledge and corrected long-held errors. For example, the age of the universe has been revised downward in part because of HST observations and now seems to be little more than twice or at most three times the 4.6 billion year age of our solar system. This has profound implications for cosmology, the study of the origin of the universe.

Five years after the HST was placed in orbit, more than 60 percent of all U.S. astronomers and astrophysicists were using HST data in their research, data made available through the Space Telescope Science Institute (STSI)

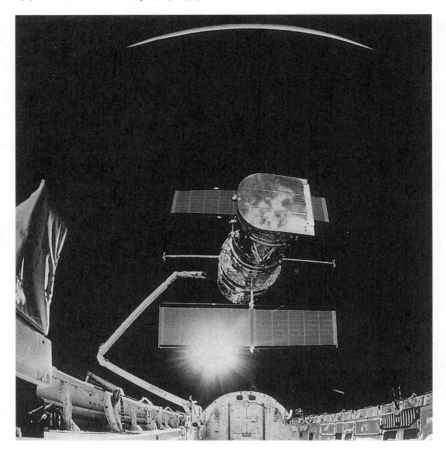

FIGURE 7.3. Photograph of the Hubble space telescope in orbit, just after being released from the space shuttle.

Source: Courtesy NASA.

located on the campus of the Johns Hopkins University, in Baltimore, Maryland. The space administration had put out a request to universities and national laboratories in the early 1980s for proposals to develop a ground-based institute to serve as headquarters for the analysis and dissemination of the data from the HST. Located near NASA's Goddard Space Flight Center, Johns Hopkins outbid the competition of other multimillion dollar proposals and won the opportunity to create the STSI.

Another satellite equipped for scientific observations was the Compton

Gamma-Ray Observatory (CGRO), launched on space shuttle *Atlantis* in April 1991 and named for Arthur H. Compton, a Nobel laureate for his pioneering X-ray studies. The CGRO carries (it is still in orbit) four scientific instruments that study the highest energy electromagnetic radiations observed in space: X-rays and gamma rays. Like the HST, the CGRO is an Earth satellite and has a planned mission duration of five to ten years. It can be reprogrammed from Earth to focus on selected stellar objects. At the end of 1995 the CGRO had observed more than fourteen hundred intense, short-lived gamma-ray bursts distributed over the entire sky. These bursts still have no completely adequate explanation and are consequently much studied. Of similar interest to astronomers and astrophysicists are the CGRO observations of especially powerful galaxies at the visible limit of the satellites. Many of these galaxies and clusters of galaxies have active nuclei— very intense emitting hot spots—at their centers. The active galactic nuclei (AGN), as they are called, are thought by most astronomers to be powered by extremely massive black holes at their cores. These attract mass from outside the black hole radii, converting the potential energy of the falling mass to kinetic energy that supplies the power of these most luminous of all known stellar bodies.

A third NASA science satellite, the cosmic background explorer (COBE), was launched earlier than the HST and CGRO, in November 1989, only a year after shuttle flights were resumed. According to the current theory of the origin of the universe, there occurred an explosion of extraordinary energy (the big bang, so-called) from which emerged the elementary particles that are the constituents of all matter and energy everywhere. Among the particles rushing away from the explosion—which account for the concept of the expanding universe—were particles of light (now called photons) that soon thereafter ceased to interact with other matter and energy as the distance separating them grew. In accord with relativity theory, the energy of the photons decreased as the universe expanded during the next ten to fifteen billion years. By now this sea of photons, which fills all space and is known as the cosmic background radiation (CBR), is radiation of very low energy or, equivalently, very low temperature. Initial observations of CBR in 1965 won a Nobel Prize for its discoverers. Early measurements by the COBE verified the existence of the CBR and further demonstrated that the temperature of the photon sea is consistent with the big bang theory.

The presence or absence of variations in the CBR temperature from point to point in space might reveal additional features of the big bang and the period immediately ensuing. These variations were what the COBE set out to measure; its success in doing so has given rise to a multitude of ground-based and balloon-flight experiments, as well as plans for a second, more sophisticated COBE satellite. The COBE data are a milestone in the development of a theory of the universe.

The HST, CGRO, and COBE science satellites have revolutionized the study of astronomy, astrophysics, and cosmology in universities throughout the world. They marked a change of emphasis within NASA from the manned-flight space agency of the 1960s and 1970s to the science and technology agency of the 1990s, a remarkable transition for any institution to make, much less a government agency. Today, NASA appears to have a bright future, both for the agency itself and for the university scientists whose research is intimately bound up with it.

Caught up in the national health care dilemma, the NIH fell further behind in funding approved proposals.

The NIH continued to struggle with its own success in the era of biomedical advances in molecular biology and gene splicing. The growth of funding in the 1980s did in fact help to sustain the promised annual rate of more than five thousand new and renewed investigator-initiated projects and ten

FIGURE 7.4. *Opposite page top*: Schematic outline of the radiochemical solar neutrino telescope, one mile (1600 meters) underground in the Homestake Gold Mine in Lead, South Dakota. The tank holds one hundred thousand gallons of perchlorethylene (a dry-cleaning fluid), which is both the target and the detector of the solar neutrinos. The auxiliary equipment is for flushing helium gas through the perchlorethylene to remove the radioactive argon atoms produced by the solar neutrinos interacting with it and for counting the individual argon atoms. This equipment began collecting data in 1967 and continues to do so today.

Source: Diagram courtesy of Raymond Davis Jr.

Opposite page bottom: Photograph of the very large array (VLA) of radio telescopes at the National Radio Astronomy Observatory in Socorro, New Mexico. The array contains twenty-seven telescopes, each twenty-five meters in diameter, located along the three legs of a Y, all of which can be pointed in the same direction. Only nine of the telescopes can be clearly seen here.

Source: Photograph courtesy of Douglas Johnson, 1981.

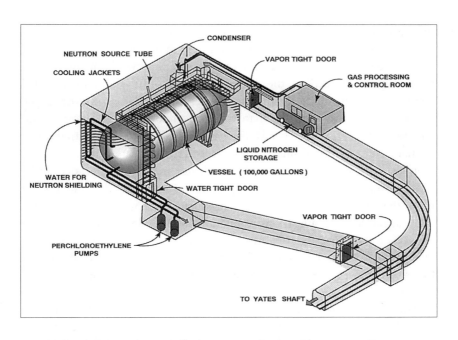

CONDENSER

NEUTRON SOURCE TUBE

VAPOR TIGHT DOOR

COOLING JACKETS

GAS PROCESSING
& CONTROL ROOM

WATER FOR
NEUTRON SHIELDING

LIQUID NITROGEN
STORAGE

VESSEL (100,000 GALLONS)

WATER TIGHT DOOR

VAPOR TIGHT DOOR

PERCHLOROETHYLENE
PUMPS

TO YATES SHAFT

thousand trainees. During that period, the portion of the NIH budget devoted to basic research increased from 52 to 63 percent.

But the number of applications, the mass of paperwork, the number of peer review study sections, the growing demands of tutorials and amended applications, and, finally, the monitoring of project progress, imposed more of an administrative burden than the NIH could manage. Tutorials and amended applications had been established to help unsuccessful applicants, particularly those unfamiliar with the research application procedure. By 1990 only 22 percent of approved investigator-initiated projects could be funded, and the backlog of approved but unfunded applications rose to more than eleven thousand. As a consequence, the number of amended applications and reapplications proliferated and tended to crowd out new ones. Even for the best young investigators, the situation was forbidding. It was no wonder that the NIH award system in general and peer review in particular were again the subjects of serious reevaluation during the decade 1985–1995.

The NIH had grown in 1995 to seven institutes concerned with almost all of medical science: heart, lung, and blood; arthritis and metabolic diseases; mental health; general medical science; neurological diseases and blindness; cancer; and allergy and infectious diseases. Neither this growth nor the progress of medical practice and biomedical research could have been foreseen in the early days when the medical community was determined to keep the NIH separate from the other federal research agencies and the government itself. The virtually independent NIH prospered through successive administrations, however, and this situation prevailed throughout the first decades of the NIH. But as the cost to the government of medical care and the NIH began rising more and more rapidly, presidents from Nixon to Clinton looked for ways to bring health care and the NIH under better financial control. Counter to this, the success of biomedical research, largely sponsored by the NIH, demanded that funding for cutting-edge research be increased. The stalemate in health care legislation that subsequently developed is more familiar than the stalemate in attempts to solve the fiscal problem of the NIH. Both situations are similar, however, in that one administration after another has been convinced that future resources will not allow the cost of either medical practice or medical research to grow at their previous rates. The general problem was illustrated in miniature in the NIH as the rising number of applications for funding and their increased complexity and cost taxed the NIH to the limit of its capability.

Streamlining the application and award procedures, or modest increases of staff, or even modest funding increases were likely to be temporary stop-gap measures in the face of the urgent demands of biomedical research. The fundamental problem before the NIH—generated by its remarkable success—would need to be addressed more generally to find a solution that could sustain reasonable, steady growth and stability.

The crisis of the award system within the NIH is especially relevant to the evolution of the science establishment. The difficulties encountered by the NIH award system are important in the continuing relationship of science and government because they raise a key question for both: at what level of funding does government say "enough" to a successful science agency that has provided the scientific basis for superior benefits to its citizens? The ongoing national debate on the general subject of health care has indicated that the question is only one aspect of government support of health care, from research laboratory to doctor's waiting room. In time, the question of "enough" will be asked of all federal science funding agencies as science expands and the allure of science continues to beckon many of the brightest and most dedicated in each new generation. For obvious reasons, the issue facing the NIH is simply the first to force the question.

The National Science Foundation remained constant to its primary function of funding basic science in diverse areas.

The NSF flirted once more with "the applied" and "the relevant" during the period from 1984 to 1990, when its director was the first to come from industry and the first to serve a full six-year term since 1969. Even so, the agency maintained its original emphasis on mathematics, science, and technology, areas that had always been the source of its strength. As a result, the NSF marched more or less sedately through the decade 1985–1995, expanding its core interests and branching carefully into new areas.

The NSF took over the funding of ground-based astronomy and low-temperature physics. It supported engineering and materials research in universities. It developed an extensive fleet of research vessels for oceanographic studies. It maintained the Antarctic research station and became a mainstay of atmospheric sciences research.

On average, about 11 percent of the roughly 135,000 science and engineering faculty in the United States applied to the NSF each year. In the ban-

Table 7.1 Science Advisers to the Presidents of the United States

ADVISER	PRESIDENT	DATES
Vannevar Bush	Roosevelt	1939–1951
Oliver Buckley	Truman	1951–1953
Lee A. Dubridge	Truman	1953–1955
Isadore I. Rabi	Truman	1955–1957
James R. Killian	Eisenhower	1957–1959
George B. Kistiakowsky	Eisenhower	1959–1961
Jerome Wiesner	Kennedy	1961–1963
Donald Hornig	Johnson	1964–1969
Lee A. Dubridge	Nixon	1969–1970
Edward E. David	Nixon	1971–1973
Guyford Stever	Nixon	1973–1974
Guyford Stever	Ford	1974–1977
Frank Press	Carter	1977–1981
George Keyworth	Reagan	1982–1987
William Graham	Reagan	1987–1989
D. Allan Bromley	Bush	1989–1993
John H. Gibbons	Clinton	1993–1999
Neal Lane	Clinton	1999–

ner fiscal year 1988, the NSF received over twenty-four thousand proposals of all kinds, which were evaluated by fifty-six thousand reviewers. The magnitude of the paperwork was enormous, well over two million pages in that year, but the NSF maintained its high efficiency by using such a large number of reviewers. The burden carried by an NSF reviewer was small compared with that carried by an NIH reviewer (the NIH's Division of Research Grants had roughly twenty-five hundred reviewers available at any time in 1987), which may have accounted for the ease with which the NSF was able to solicit reviews.

The NSF distributed annually close to $2 billion in support of basic science and technology. Nevertheless, the problem of making ends meet, of unwittingly promising more than it can deliver, is faced today by the NSF, just as it is by the NIH. The root cause is very much the same in both agencies: the demand for funds exceeds the supply by an amount that cannot be reduced by means that treat only the symptoms of the problem. The recurrent question before the government is deceptively simple: how much support is enough? Unfortunately, the answer is not nearly so simple; indeed, there are many questions to be considered before the question of "enough"

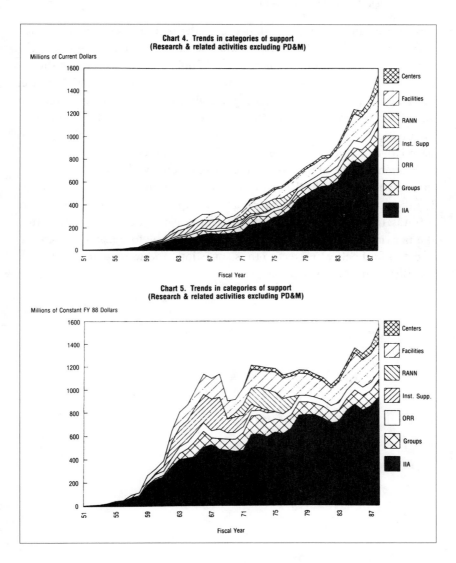

FIGURE 7.5. The NSF budget showing categories of support in research and related activities 1951–1987: RANN, research applied to national needs; Inst. Supp., institutional support; ORR, other research resources; Groups, group research; IIA, individual investigator awards; PD&M, program development and management. *Top*: current dollars; *bottom*; FY 1988 dollars.

Source: T. N. Cooley and Deh-I Hsiung, *Funding Trends and Balance of Activities: National Science Foundation, 1951–1988*, NSF 88-3 (Washington, D.C.: National Science Foundation, 1988), p. 7.

can even be addressed. For instance, considering the investments of time, energy, and taxpayer money that go into the education and training of a scientist, is it efficient to fund the research of only about 20 percent of them? Does that funding level indicate that there are too many scientists? Is it wise to send a message that discourages young people from acquiring an advanced university degree in science? And, if so, how should that be done? Finally, how large should the yearly investment in the science establishment be, relative to other government commitments and expenditures?

Even in the case of the NIH, where the research is without doubt directed toward national needs, the answers to those questions do not come easily. For the NSF, which is more diverse and less applied in much of the science it funds, the answers are harder to determine. Nevertheless, fifty years after the marriage of science and government, it is time to make a thoughtful effort to find better answers to those questions and to prepare the science establishment for the next half century.

CHAPTER 8

The Future: 2000 and More

The science establishment that emerged in the United States during the past half century is a phenomenon in its own right: a unique partnership of government and private institutions, of administrators and scientists and engineers, brooded over by the Congress, which finances it, and pretty much allowed to go its own way under the benign neglect of the White House.

Much of the success of the enterprise, particularly in the early years, was due to the informal, personal way in which the federal science agencies conducted their business. The agencies were staffed by administrators who had had careers as scientists and who saw themselves as brokers between their scientific colleagues and the government. The peer review system in which other active scientists were asked for value judgments on proposed research also contributed to a family atmosphere in each of the agencies.

In the mid-1950s, for example, the AEC offered to finance construction by universities of several high-energy particle accelerators in addition to those it had sponsored soon after WWII. The idea was to encourage two or perhaps three universities in the same geographical region to collaborate on the construction and subsequent operation of an expensive experimental facility for faculty research and student training. Use of the facility would be open to faculty at other universities, if their proposals passed muster by reviewers, but responsibility for the design, construction, and operation of the facility would rest with the governing universities. Princeton University and the University of Pennsylvania, located less than an hour apart by train, tried to form such a collaboration, and two physics faculty representatives from each institution began regular meetings to iron out the details.

That was not so easy. The Penn people thought a machine that accelerated electrons, with which they had some experience, was more promising for future research in elementary particle physics. The Princeton people saw the future in terms of a proton accelerator, which, at the time, was less risky to build and to which one of its representatives—a former student of Ernest Lawrence at Berkeley—brought direct design and construction experience. The disagreement persisted for the better part of a year, never turning ugly because the other Princeton representative was Henry Smyth, the former AEC commissioner, whose vast federal experience in disputes kept the meetings civil and on track.

While still far apart on the specific technical nature of the facility, how it would be administered, and its location, the four negotiators were asked to go to Washington to report to the director of energy research of the AEC, a well-known physicist named Tom Johnson. In Johnson's office the conversation went something like this:

JOHNSON: Well, I'm glad to see that you are actively collaborating at last. How far have you progressed with the design and location of the facility?

SMYTH (AS SPOKESPERSON FOR THE NEGOTIATORS): Tom, we're sorry to say that we are unable to agree on the design or the location, although we have struggled mightily to overcome our differences.

JOHNSON: That's too bad, really too bad! Here in the bottom drawer of my desk (reaching down and pulling the drawer open) I have five million dollars to give you now if you can agree to collaborate on a facility. But if you can't, I am prevented from giving anything to either institution acting alone.

(There was a long pause during which the university representatives surreptitiously eyed one another and waited for Smyth to respond. Instead, Johnson retook the initiative.)

JOHNSON: Now with that in mind, I suggest you go into the office next door to this one, which happens to be unoccupied, and see if you can't find a compromise that will make positive use of the five million dollars.

(With much nodding of heads, the four visitors went next door and returned to Johnson less than five minutes later, when Smyth told him that the two universities would collaborate on a proton accelerator, for which the design was already prepared, to be located in the James Forrestal Research Area in Princeton, and with each university sharing equally in the administrative and scientific responsibility.)

That was the origin of the Princeton-Pennsylvania Accelerator. It attracted young faculty and students to both universities and had a valuable productive life until it was outmoded a decade later by construction of a much more energetic proton accelerator at Fermilab. Another high-energy accelerator, the Cambridge Electron Accelerator, was built by the two-university collaboration of Harvard and MIT under the same AEC support scheme. Scientists in other fields could tell similar stories about their experiences with the NIH and the NSF.

To their credit, all the federal science agencies have tried to preserve close, personal relationships with the scientists they support. But the rate of growth of the science establishment has made the task tremendously difficult. Over the years, the pressure to centralize and enlarge administrative functions to facilitate accountability to Congress has been resisted but not completely avoided. The future growth in breadth and depth of science will exert ever-greater pressure on the science agencies to become less personal and to streamline activities. The pressure is sure to increase as the nation adjusts to the end of the cold war, responds to new international commitments, and seeks to improve the quality of life and the health and long-term security of its citizens.

Imposition of limits on the flexibility of the science establishment, even when done with the lightest hand by the government, has been shown to be counterproductive. Nevertheless, the motivation to try again will be hard to resist. In time, therefore, the burden of guidance of the science establishment will rest more heavily on the scientists themselves than it does at present. They are most likely to be aware of the illusory advantages of increased centralization and of the need to uphold the primacy of the active researcher within the framework of the science agencies. Scientists in concert with science agencies have the knowledge and the firsthand experience to furnish advice that will bring about beneficial changes in the operation of the science establishment.

One way in which scientists might assume greater responsibility for their future and the future of the science establishment would be to participate more actively in formulating science agency budgets and agency recommendations to Congress.

Scientists and the Science Budget.

The U.S. science budget consists of many loosely defined categories that make up a substantial part of the so-called discretionary funds that in turn constitute roughly 16 percent of the total federal budget. Today, there are about twenty federally funded agencies whose missions involve research and technology. Four of those agencies supply almost all the federal funding for scientific research in colleges and universities.

The annual federal research and development budget is about $70 billion, of which roughly one-half is spent on functions that, while important, do not create new knowledge or develop new technology. Examples include evaluation of new aircraft and new weapon systems at the Department of Defense (DOD), nuclear weapons work at the DOE, and mission evaluation at NASA. The remainder, roughly $35 billion, is spent on basic science and technology, roughly as follows: the Department of Health and Human Services (DHHS), in which the NIH resides, spends about 30 percent; NASA, 15 percent; the DOE, 14 percent; and the NSF, 5 percent. Science and technology sponsored or done by the DOD accounts for 22 percent. Other federal agencies concerned with agriculture, the environment, and transportation, whose science and technology are peripheral, spend 14 percent. This distribution of science and technology funds indicates the situation during a multiyear period. Changes in the relative or total amounts occur slowly, usually in response to a stimulus from outside the science establishment.[1]

One question that might be asked of the science establishment is whether the pattern of fund sharing and the manner in which the pattern is shaped are adequate to respond to recent world changes and to the new economic and social climate in the United States. For example, the NSF spends all but about one-fifth of its 5 percent share of the science and tech-

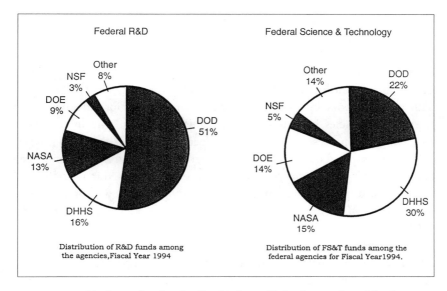

Federal R&D

Other 8%
NSF 3%
DOE 9%
NASA 13%
DOD 51%
DHHS 16%

Distribution of R&D funds among
the agencies, Fiscal Year 1994

Federal Science & Technology

Other 14%
DOD 22%
NSF 5%
DOE 14%
DHHS 30%
NASA 15%

Distribution of FS&T funds among the
federal agencies for Fiscal Year 1994.

FIGURE 8.1. Pie charts showing the distributions of federal research and development (FR&D) funds (*left*) and federal science and technology (FS&T) funds as of 1995. Roughly one-half of the R&D budget includes important functions that generally do not create new knowledge or develop new technology.

Source: National Academy of Sciences Committee on Criteria for Federal Support of Research and Development, *Allocating Federal Funds for Science and Technology* (Washington, D.C.: National Academy Press, 1995), p. 137.

nology budget on the basic research of individual investigators. Envision the effect of an increase in that share from 5 to 7 percent, an increase obtained by very small decreases in the shares of the other agencies. The substantial fractional increase in the NSF's funding capability would allow as much as a 50 percent increase in the number of researchers it supports, particularly many gifted young investigators who otherwise find it so difficult to obtain support for their work. Compare that major benefit to the minor losses of the other agencies.

Concern with sharpening the focus of responsibility is, however, more general than the question of how much to take from one agency to give to another. It concerns, rather, how to assess the system by which allocations are made and how to modify the system to improve the components of the science establishment itself. To do this, the university science communities should be given and accept greater responsibility for the process by which their science agency budgets are generated.

In our system of governance, any substantive change in the operation of the science establishment needs to be enacted by Congress. But the specifics of the change and the impetus for it should involve university scientists. This might be done by enlarging their present role as informal advisers to the science agencies, including them as regularized consultants in the agencies' budgetary processes. At present, an agency may ask scientists to review a research proposal or a report on the progress of a research program. The agency's request and each scientist's response are submitted in writing, but the procedure is informal, limited to that advisory review function. The agencies accord scientists little voice in their science policies and no formal status.

It may seem foolish to suggest that university scientists, the principal beneficiaries of government funding for science, be assigned greater responsibility for the policies of the science agencies in general and their budgets in particular. Nevertheless, scientists are most likely to know what improvements are necessary and how they might be implemented. Moreover, the immediacy and closeness of their involvement will ensure continued participation, while diversity and self-interest will protect against excessive selfishness by any specific discipline.

Scientists and engineers are asked to review not only programs under consideration by an agency for inclusion in its budget; they are also asked by professional journals to review manuscripts submitted for publication. And universities and national laboratories make similar requests with regard to appointments and promotions of individuals. In the main, most reviewers do a good job reviewing manuscripts and evaluating individuals for appointments and promotions. It is principally when evaluating programs and facilities—new and old—that they perform inadequately and often adversely influence the budget process. The influence is not always short term because decisions on programs and facilities usually involve commitments that last for a number of years.

If a review concerns an older program or facility, factors other than scientific merit and accomplishment, such as an institution's past reputation, may affect a funding agency's final decision with a degree of importance not given by reviewers. Moreover, discussion to promote understanding of agencies' decisions rarely takes place. In renewals of many modestly funded programs and facilities, reviewers are usually unaware of the identities and opinions of other reviewers—unlike the legal jury system—and learn of the final decision of the agency only as it seeps down through the system as an accomplished fact.

If a proposed program or facility is new and expensive, the prospective funding agency forms a temporary review committee to which it gives detailed instructions. Those instructions, and not primarily scientific merit, occasionally determine the fate of the proposal. Neither this nor any other statement I have made is meant to question the integrity of the individuals in any agency but rather to reiterate that peripheral criteria such as the geographical distribution of facilities or separation of function and emphasis between facilities may outweigh scientific merit in agencies' decisions. Nor do these comments apply equally to each of the agencies I have discussed here. In rare but significant instances, political action outside an agency may be used to circumvent the clear recommendation of an agency's external advisory committee. The reviewing committee has no formal recourse when it disagrees with an agency's decision. Reviewers serve at the discretion of the agency and have no independent avenue through which to manifest their disagreement. And frequently there is no traceable path between the scientific review and the final agency decision. Scientists on the review committee and even the proponents of a proposal that receives a negative review rarely object to decisions of this kind for want of an accepted, formal means to do so. And the habit of conforming to the final decision of an agency—based on what is advertised to be a strictly scientific review—is not easily broken.

For these reasons many scientists lack confidence in the review system when it deals with scientific programs and facilities. They question the usefulness of their limited advisory role in the decisions the agencies make. This lack of confidence often leads them to put less than a full effort into the review process. It tolerates programs and facilities that do not satisfy the high standards of the profession and rejects some that do. Scientists implicitly view the effectiveness of the present system as a secondary—rather than a primary—responsibility of theirs. Consequently, the system does not adequately provide the quality assessment necessary to sustain and advance the health of the U.S. science establishment in the coming decades.

One big step toward bringing important policy decisions and the science budget into better focus would be to charge university and industry scientists with clearly specified responsibility for the information and advice on which scientific decisions within each agency are based. This would introduce in a regularized, formal way the "essential element of outside expert judgment that is the bedrock of quality assessment in research and development."[2] It would bring scientists and the science agencies into a better working relationship. The passive role now played by scientists would be replaced by an active role that would involve them more fully and responsibly in the reasoning at the heart of the decision process. They would participate in the compromises and trade-offs that are an integral, and perhaps the most important, part of the creation of any science agency plan. Scientists would be aware of and understand the fate of their recommendations as the process evolves within the agencies. This new responsibility placed on the science communities would provide for a mutually beneficial measure of feedback and a level of collegiality among scientists and present and future policy makers. But it need not and would not intrude on the responsibility for decisions of policy makers within an agency. That responsibility would remain, as before, in the hands of agency officials. Moreover, the new role of the scientists could lead to greater confidence on the part of Congress in the agencies' requests.

The intent of this proposal is to recommend a structure analogous to the advisory committee structure that was created by a clever compromise in the Atomic Energy Act of 1946. That compromise led to a Military Liaison Committee (MLC) and a General Advisory Committee (GAC) attached to the Atomic Energy Commission (AEC), each with clearly defined duties and limitations. These committees neither attended AEC meetings nor had a vote in commission decisions, but each had a recognized, formal advisory function to perform for the commission. The AEC in turn had the responsibility to inform the committees of each of its decisions at an appropriate time. In case

FIGURE 8.2. *Opposite page top*: Comparisons of overall spending on research in the United States by private industry and the federal government. Also shown is a comparison of the sources of scientific papers cited on U.S. industrial patents in 1993–1994.

Source: William J. Broad, *New York Times*, May 13, 1997, p. C1.

Opposite page bottom: Trends in federal discretionary and R&D outlays during 1960–1993. Total domestic discretionary outlays, total support for federal research and development, and total support for nondefense research and development are plotted as percentages of the total federal outlays (right ordinate for total and left ordinate for R&D) for the period from 1960 through 1993. The different administrations are indicated for reference.

Source: William J. Broad, *New York Times*, May 13, 1997, p. C1.

Research: Public vs. Private

Private industry in the United States has surpassed the Federal Government in overall spending on research. But a recent study has found that more than 70 percent of the scientific papers cited on U.S. industrial patents came from public science — research performed at universities, government labs and other public agencies.

Most research dollars now come from Industry . . .
In billions of current dollars

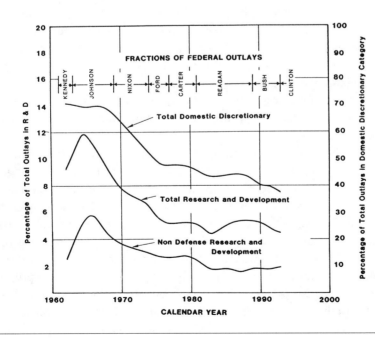

. . . but patents cite public science most often.
Source of scientific papers cited on U.S. industrial patents in 1993-1994

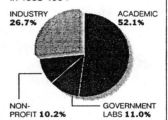

INDUSTRY **26.7%**   ACADEMIC **52.1%**

NON-PROFIT **10.2%**   GOVERNMENT LABS **11.0%**

Sources: National Science Foundation; CHI Research Inc.

The New York Times

FRACTIONS OF FEDERAL OUTLAYS

KENNEDY · JOHNSON · NIXON · FORD · CARTER · REAGAN · BUSH · CLINTON

Total Domestic Discretionary

Total Research and Development

Non Defense Research and Development

Percentage of Total Outlays in R & D

Percentage of Total Outlays in Domestic Discretionary Category

CALENDAR YEAR

1960   1970   1980   1990   2000

of a severe disagreement with a decision, the MLC—with members appointed by the armed service secretaries—could refer their objections to the secretaries for further appeal; the GAC—with members appointed by the president—could take its objections directly to the president.

In general, that structure worked well for thirty years. The AEC had the benefit of knowledgeable, mandated advisory committees but—with certain restrictions—was formally independent of them. The committees, representing different constituencies, were kept informed and moreover had well-delineated avenues through which to voice their objections when necessary.

Consider the formation of several external advisory committees, one of which would be accredited to each of the four science agencies I have featured. Each committee, consisting of between five and ten members, might be chosen from a slate recommended by relevant professional organizations. To give them appropriate status and formalize their responsibility, each committee would be appointed by and be responsible to the president's science adviser.

The terms of committee members would be for two or three years, staggered so that at least one-third of the members would be replaced each year. Each committee would choose its own chairperson, who would serve for one year. Members would receive no compensation, but their expenses would be paid. No member would participate in any deliberation relating to the institution with which he or she is affiliated. Nor would members be considered as representatives of any institution, scientific organization, or discipline. Their loyalty would be to an agency to which they were appointed. Their job would be to help that agency carry out its science functions in the most efficient, economical, and productive way.

The duties of each science advisory committee would be to consult with and advise the policy and budget makers on issues related to the agency budget, in a manner acceptable to all parties. Each committee would have no administrative function and would be concerned only with matters related to the agency budget. Each agency, in concert with its external committee, would define the way in which the two bodies would jointly operate, keeping in mind the need to safeguard against excessive partisanship.

Skepticism on three counts is likely to fuel criticism of this proposal. First, can scientists overcome personal self-interest; second, will they be satisfied with their limited responsibility; and third, will they be able to maintain the requisite confidentiality? None of these counts is justified given the past behavior of scientists or of public-spirited citizens in general, but much care will be required to ensure that none is a future pitfall.

Indeed, the peer review system now in operation partly circumvents these objections. What I propose here is not to do away with that system but to strengthen it through a well-defined, formal structure, one that would share the burden of fact-finding and the weighing of alternatives between the agencies and scientists. Decisions would rest, as before, with the officials of each agency.

The most significant benefit to come from involving scientists more deeply and formally in the operation of the science establishment might be the demand placed on them—explicitly for the first time—to explain to Congress and the public what they do and why they should be supported by the taxpayers. Scientists will need to differentiate convincingly the support of science from an entitlement and to recognize that the people who pay the bills deserve to appreciate the intellectual and spiritual excitement of the scientific enterprise as well as its practical benefits.

The Power of Science and Technology for Good and Evil.

The new century we are about to enter will be filled with change that we can only dimly identify now. But it is certain that progress in science and the interplay of science and technology will not stop in the future. It is fueled by human curiosity and the immense satisfaction we humans find in learning and knowing. Increasing numbers of each new generation will rediscover the attraction of the endless frontier of science and choose to spend their lives exploring it, which leads to a question about the science establishment in the United States of the mid-twenty-first century: Is it likely to have the same form that it has now, or will it be the oppressive, bureaucratic monster envisioned in the novels of Aldous Huxley and George Orwell?

The story of U.S. science contains more virtues than not. The importance of encouraging and supporting the science establishment has long been a vital component of American prosperity and culture. But it is prudent to be aware that the enormous strength of the science establishment could be marshaled and directed toward misguided or even inherently evil goals. Earlier chapters in this book discuss the power that has been generated by science and technology when organized on a national scale and devoted to predetermined goals. Fortunately, the weapons that were pro-

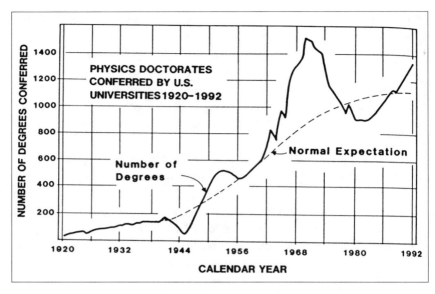

FIGURE 8.3. Physics doctorates awarded by U.S. universities, 1920–1992. The curve labeled "Normal Expectation" has been drawn in as a rough guide to what might have been a desirable expectation. It emphasizes, however, several features of these data. Those who were prevented by World War II from completing their Ph.D. education did so after the war. The rapid increase in Ph.D. production in the late 1960s came in response to the federal programs aimed at producing the trained personnel that projections at the time suggested would be required in the U.S. military and space programs. This crash program was terminated in 1968 when it was found that the goals established for 1970 had already been reached, and the resulting student disenchantment and the lack of anticipated opportunities led to the rapid decrease in production in the 1970s. Projections of coming shortages, published in the early 1980s, resulted in a new increase.

Source: D. Allan Bromley, *The President's Scientists* (New Haven: Yale University Press, 1994), p. 118. Courtesy of D. Allan Bromley.

duced in the United States during WWII were used with restraint. Even so, the soul-searching that preceded the use of atomic weapons in Japan and the subsequent debate that persists after so many years highlight the vital difference between an ethical government and a government corrupted by its own power. What a different turn the world would have taken if Nazi Germany had successfully created an atomic bomb before the Allies did.

In the fifty years since WWII, the power of organized, directed science and technology has increased many times. Advances in all phases of weaponry produced by scientists and engineers in the Departments of Energy and Defense, acting together with industry, are evidence of what can be accomplished.

A case in point is the intercontinental missile defense system known as Star Wars. As proposed by President Reagan in the 1980s, at the height of the cold war, the plan took many conceptual forms, which included satellite-launched interceptors with laser weaponry, Earth-based long-range missiles, and various combinations of them. Most scientists outside the government viewed the project as ill conceived and unfeasible. Indeed, it never deployed a defense system or even showed significant progress toward one. But neither did it die. Today there is a scaled-down version of the effort, the National Missile Defense, with a budget still in the billions, even though the consensus of the science community with regard to its promise remains negative.

We cannot easily call into question the existence of the military-industrial complex—so named by President Eisenhower—that makes possible and promotes such efforts, because that complex also engages successfully in other enterprises on which our national security rests. But that is all the more reason why we need to oversee its behavior with great care.

The line that separates science and technology organized for good from science and technology organized for poorly conceived or even inherently evil goals has never been sharp and is likely to become less so with time. Consider, for example, Earth-orbiting satellites equipped with high-resolution, digitized cameras feeding powerful computers that allow us to inspect and tally minute features and resources of our planet. This is a powerful instrument for good. But the same instrument serves also to monitor the hour-by-hour behavior of farms, businesses, and armies and, quite possibly, the smallest details of individual lives. Everything can be observed and studied for possible activity that might be described as injurious to world peace or the stability of the state. The capability exists to do this with an efficiency far superior to the spine-chilling but clumsy efforts of the Soviet secret police to control its fellow citizens during the cold war.

We know of only one way to protect against the evils that follow from the misuse of organized science and technology. That way lies through the fullest exchange of ideas and opinions among all segments of society and particularly between governments and scientists. In that way, any serious threat will be evident and can possibly be overcome. There will always be citizens in all walks of life who will not be taken in by specious arguments to justify the use of science and technology for questionable national goals. Scientists, however, are best equipped to recognize and expose the weaknesses in those arguments. They can defend most effectively against the abuse of their own work. Scientists constitute a vital element in the protective armor of a nation against such misadventure. Here again is reason for the inclusion of scientists within the science agencies. In their ceaseless search for the truth, scientists enhance the quality of life. Equally important, they serve to safeguard that quality and the integrity of science from those who would pervert them.

2. Love at First Sight: 1939–1945

1. James Phinney Baxter III, *Scientists Against Time* (Boston: Little, Brown, 1946), p. 138.
2. Ibid., p. 236.
3. R. G. Hewlett and O. E. Anderson Jr., *A History of the U.S. Atomic Energy Commission*, vol. 1, *1939/1946* (Berkeley: University of California Press, 1989), p. 20.
4. Ibid., p. 47.
5. Ibid., p. 394.
6. Ibid., p. 396.
7. Ibid., p. 415.
8. Jerome B. Wiesner, *Biographical Memoir of Vannevar Bush* (Washington, D.C.: National Academy of Sciences, 1979), p. 89.
9. Vannevar Bush, *Pieces of the Action* (New York: Morrow, 1970), p. 26.
10. Ibid., p. 302.

3. Courtship: 1945–1955

1. R. G. Hewlett and O. E. Anderson Jr., *A History of the U.S. Atomic Energy Commission*, vol. 1, *1939/1946* (Berkeley: University of California Press, 1989), p. 505.

2. R. G. Hewlett and J. M. Holl, *A History of the U.S. Atomic Energy Commission*, vol. 3, *1953/1961* (Berkeley: University of California Press, 1989), p. 99.

3. Richard Mandel, *A Half Century of Peer Review (1946–1996)* (Alexandria, Va.: Division of Research Grants, National Institutes of Health, Logistic Applications, 1996), p. 21.

4. Ibid., p. 39.

5. Ibid., p. 65.

4. Marriage: 1955–1965

1. R. E. Bilstein, *Orders of Magnitude: A History of the NACA and NASA, 1915–1990*, NASA SP-4406 (Washington, D.C.: NASA, 1990), p. 59.

2. R. G. Hewlett and J. M. Holl, *Atoms for Peace and War. 1953–1961: Eisenhower and the Atomic Energy Commission (A History of the Atomic Energy Commission, vol. 3)* (Berkeley: University of California Press, 1989), p. 495.

3. Ibid., p. 209.

4. Ibid., p. 537.

5. Richard Mandel, *A Half Century of Peer Review (1946–1996)* (Alexandria, Va.: Division of Research Grants, National Institutes of Health, Logistic Applications, 1996), p. 96.

5. End of the Honeymoon: 1965–1975

1. Richard Mandel, *A Half Century of Peer Review (1946–1996)* (Alexandria, Va.: Division of Research Grants, National Institutes of Health, Logistic Applications, 1996), p. 147.

2. Ibid., p. 148.

6. Estrangement and Reconciliation: 1975–1985

1. T. R. Fehner and Jack M. Holl, *Department of Energy, 1977–1994: A Summary History* (Oak Ridge, Tenn.: Office of Scientific and Technical Information, 1995), p. 22.

2. Ibid., p. 26.

3. Ibid.

4. Ibid., p. 27.

5. Ibid., p. 30.

6. Ibid., p. 31.

7. G. T. Mazuzan, *Good Science Gets Funded: The Historical Evolution of Grant Making at the National Science Foundation* (Washington, D.C.: National Science Foundation, 1992), p. 77.

8. Ibid., p. 78.
9. Ibid., p. 79.
10. Ibid.

11. Richard Mandel, *A Half Century of Peer Review (1946–1996)* (Alexandria, Va.: Division of Research Grants, National Institutes of Health, Logistic Applications, 1996), p. 153.

12. Ibid., p. 179.

7. Golden Anniversary: 1985–1995

1. T. R. Fehner and Jack M. Holl, *Department of Energy, 1977–1994: A Summary History* (Oak Ridge, Tenn.: Office of Scientific and Technical Information, 1995), p. 60.

2. Ibid., p. 47.
3. Ibid., p. 75.
4. Ibid., p. 76.
5. Ibid., p. 88.
6. Ibid.
7. Ibid.
8. Ibid., p. 46.

8. The Future: 2000 and More

1. The distinction between the two halves of the R&D budget and the budget figures cited in the text and the charts in figure 8.1 are from National Academy of Sciences Committee on Criteria for Federal Support of Research and Development, *Allocating Federal Funds for Science and Technology* (Washington, D.C.: National Academy Press, 1995). There have been new federal science budgets in the interval between 1995 and the present, but these do not significantly modify the general content of this chapter. For example, the Clinton administration's spending figures for FY 2000 indicate that R&D outlays would come to $73.6 billion, about 0.5 percent more than the projected expenditures in FY 1999; see *Physics Today* 52, no. 4 (April 1999): 56.

2. National Academy of Sciences, *Allocating Federal Funds*, p. 27.

References

Acheson, Dean. *Present at the Creation*. New York: Norton, 1969.

Ad Hoc Working Group on Research Intensive Universities and the Federal Government. *Report*. Washington, D.C.: U.S. Government Printing Office, 1992.

Association of American Medical Colleges. "A Policy for Biomedical Research." *Journal of Medical Education* 46 (1971).

Baxter, James Phinney III. *Scientists Against Time*. Boston: Little, Brown, 1946.

Berkner, L. V. "Wanted: A National Science Policy." *Atlantic Monthly* 201, no. 40 (1958).

Bilstein, R. E. *Orders of Magnitude: A History of the NACA and NASA, 1915–1990*. NASA SP-4406. Washington, D.C.: NASA, 1990.

Brode, W. R. "Development of a Science Policy." *Science* 131, no. 9 (1960).

Bromley, D. Allan. *The President's Scientists*. New Haven: Yale University Press, 1994.

Brooks, Harvey. *The Government of Science*. Cambridge: MIT Press, 1968.

Bush, Vannevar. *Modern Arms and Free Men*. New York: Simon and Schuster, 1949.

——. *Pieces of the Action*. New York: Morrow, 1970.

——. *Science: The Endless Frontier*. Washington, D.C.: U.S. Government Printing Office, 1945.

——. *Science Is Not Enough*. New York: Morrow, 1967.

Chayes, Abram, and Jerome B. Wiesner, eds. *ABM: An Evaluation of the Decision to Deploy an Antiballistic Missile System.* New York: New American Library, 1969.

Committee on Government Operations, U.S. Senate, 87th Cong., 2d sess. *Create a Commission on Science and Technology?* Washington, D.C.: U.S. Government Printing Office, 1962.

Committee on Science and Public Policy of the National Academy of Sciences. *Federal Support of Basic Research in Institutions of Higher Education.* Washington, D.C.: National Academy Press, 1964.

Conant, James B. *Modern Science and Modern Man.* New York: Columbia University Press, 1952.

Cooley, T. N., and Deh-I Hsiung. *Funding Trends and Balance of Activities: National Science Foundation, 1951–1988.* NSF 88-3. Washington, D.C.: National Science Foundation, 1988.

Dupree, A. Hunter. *Science in the Federal Government: A History of Policies and Activities.* Cambridge: Harvard University Press, Belknap Press, 1957.

Fehner, T. R., and Jack M. Holl. *Department of Energy, 1977–1994: A Summary History.* Oak Ridge, Tenn.: Office of Scientific and Technical Information, 1995.

Garber, S. J. *Research in NASA History.* Washington, D.C.: NASA, 1997.

Golden, W. T., ed. *Science Advice to the President.* 2d ed. Washington, D.C.: AAAS Press, 1993.

Greenberg, Daniel S. *The Politics of Pure Science* New York: New American Library, 1967.

Groves, Leslie R. *Now It Can Be Told: The Story of the Manhattan Project.* New York: Da Capo, 1962.

Hewlett, R. G., and O. E. Anderson Jr. *The New World: A History of the U.S. Atomic Energy Commission.* Vol. 1, *1939/1946.* Washington, D.C.: U.S. Atomic Energy Commission, 1972.

Hewlett, R. G., and Francis Duncan. *Atomic Shield: A History of the U.S. Atomic Energy Commission.* Vol. 2, *1947/1952.* Washington, D.C.: U.S. Atomic Energy Commission, 1972.

Hewlett, R. G., and J. M. Holl. *Atoms for Peace and War.* Vol. 3, *1953–1961: Eisenhower and the Atomic Energy Commission.* Berkeley: University of California Press, 1989.

Kevles, D. J. "Scientists, the Military, and Control of Postwar Defense Research: The Case of the Research Board for National Security, 1944–46." *Technology and Culture* 16, no. 20 (1975).

———. "The National Science Foundation and the Debate Over Postwar Research Policy, 1942–45." *Isis* 68, no. 5 (1977).

Kevles, D. J., and L. Hood, eds. *The Code of Codes.* Cambridge: Harvard University Press, 1992.

Mandel, Richard. *A Half Century of Peer Review (1946–1996).* Alexandria, Va.: Division of Research Grants, National Institutes of Health, Logistic Applications, 1996.

Mazuzan, G. T. *A Brief History of the National Science Foundation, 1972–1985.* Washington, D.C.: National Science Foundation, 1987.

———. *Good Science Gets Funded: The Historical Evolution of Grant Making at the National Science Foundation.* Washington, D.C.: National Science Foundation, 1992.

Meadows, D. H., D. L. Meadows, J. Randers, and W. W. Behrens III. *The Limits to Growth.* New York: Universe, 1972.

National Academy of Sciences Committee on Criteria for Federal Support of Research and Development. *Allocating Federal Funds for Science and Technology.* Washington, D.C.: National Academy Press, 1995.

National Aeronautics and Space Administration. *Aeronautics and Space Report of the President: Fiscal Year 1995 Activities.* Washington, D.C.: NASA, 1993.

National Institutes of Health Office of Program Planning and Evaluation. *Issue Paper on the Training Programs of the NIH, Part I.* Washington, D.C.: National Institutes of Health, 1970.

National Science Foundation. *Investing in Scientific Progress.* NSF-61-27. Washington, D.C.: National Science Foundation, 1961.

———. *Research and Development and the Gross National Product, Review of Data on R&D.* No. 26. NSF-61-9. Washington, D.C.: National Science Foundation, 1961.

National Science Board. *The National Science Board and the Formulation of National Science Policy.* NSB-81-440. Washington, D.C.: National Science Board, 1981.

———. *Science at the Bicentennial: A Report from the Research Community.* Washington, D.C.: U.S. Government Printing Office, 1976.

Okun, L. B. *Particle Physics: The Quest for the Substance of Substance.* Trans. V. I. Kisin. New York: Harwood Academic Publishers, 1985.

Patel, C. Kumar N., ed. *Reinventing the Research University.* Los Angeles: UCLA Publication Design Services, 1995.

President's Commission on National Goals. Science section of *A Great Age for Science—Goals for Americans.* Englewood Cliffs, N.J.: American Assembly, Prentice-Hall, 1960.

President's Council of Advisors on Science and Technology. *Reports.* Washington, D.C.: U.S. Government Printing Office, 1992.

Rhodes, Richard. *Making of the Atomic Bomb.* New York: Simon and Schuster, 1986.

———. *Nuclear Renewal.* New York: Whittle, 1993.

"Rising Outlays for Research and Development." *U.S. News and World Report,* April 3, 1961, p. 24.

Rivlin, Alice M. *The Role of the Federal Government in Financing Higher Education.* Washington, D.C.: Brookings Institution, 1961.

Sakharov, Andrei. *Memoirs*. Trans. Richard Lourie. New York: Knopf, 1990.

Sapolsky, Harvey M. *Science and the Navy: The History of the Office of Naval Research*. Princeton: Princeton University Press, 1990.

Seaborg, Glenn (chairman of the President's Science Advisory Committee). *Scientific Progress*. Washington, D.C.: U.S. Government Printing Office, 1961.

Serber, Robert. *Peace and War*. New York: Columbia University Press, 1998.

Shannon, J. A. "The Advancement of Medical Research: A Twenty-Year View of the Role of the NIH." *Journal of Medical Education* 42, no. 97 (1967).

Shannon, J. A., ed. *Science and the Evolution of Public Policy*. New York: Rockefeller University Press, 1973.

Smith, Bruce L. R. *American Science Policy Since World War II*. Washington, D.C.: Brookings Institution, 1990.

Smyth, Henry D. *Atomic Energy for Military Purposes: A General Account of the Scientific Research and Technical Development that Went into the Making of Atomic Bombs*. Princeton: Princeton University Press, 1945.

Stewart, Irvin. *Organizing Scientific Research for War; The Administrative History of the OSRD*. Boston: Little, Brown, 1948.

Stine, Jeffrey K. *A History of Science Policy in the U.S., 1940–1985*. Report prepared for the Task Force on Science Policy, Committee on Science and Technology, House of Representatives, 99th Cong., 2d sess. Washington, D.C.: U.S. Government Printing Office, 1986.

Stoff, Michael B., Jonathan F. Fanton, and R. Hal Williams. *The Manhattan Project*. Philadelphia: Temple University Press, 1991.

Ward, Barbara. *Spaceship Earth*. New York: Columbia University Press, 1966.

Waterman, A. T. "The National Science Foundation: A Ten Year Resume." *Science* 131, no. 1341 (1960).

Whitehead, Alfred North. *Science and the Modern World*. New York: Macmillan, 1925. Reprint. New York: New American Library, Mentor Books, 1956.

Wiesner, Jerome B. *Biographical Memoir of Vannevar Bush*. Washington, D.C.: National Academy of Sciences, 1979.

Zachary, G. Pascal. *Endless Frontier: Vannevar Bush, Engineer of the American Century*. New York: Free Press, 1997.